Collins

ision

Maths

hs

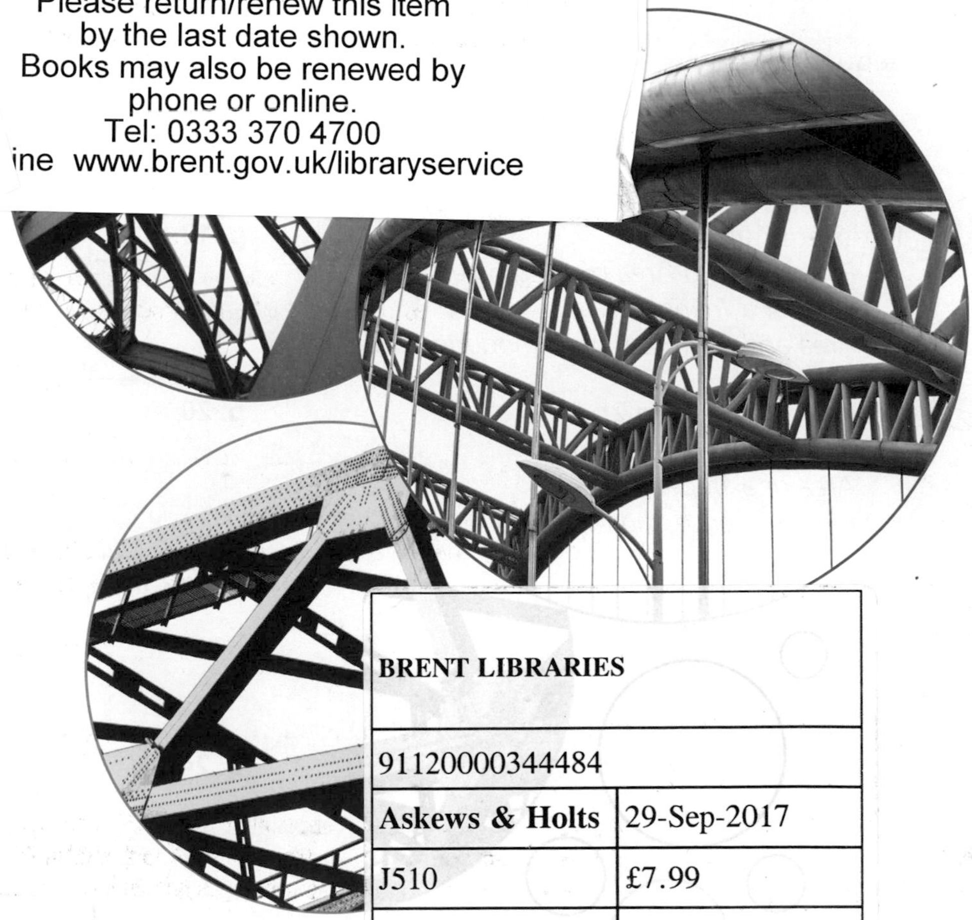

Standard

KS3

Revision Guide

bdullah, Rebecca Evans and Gillian Spragg

Contents

		Revise	Practise	Review
Key Stage 2: Key Concepts				p. 14 ☐
N Number 1 *Positive and Negative Numbers* *Multiplication and Division* *BIDMAS*		p. 6 ☐	p. 16 ☐	p. 26 ☐
N Number 2 *Squares and Square Roots* *Prime Factors* *Lowest Common Multiple and Highest Common Factor*		p. 8 ☐	p. 16 ☐	p. 26 ☐
A Sequences 1 *Function Machines* *Sequences* *Finding Missing Terms*		p. 10 ☐	p. 17 ☐	p. 27 ☐
A Sequences 2 *The nth Term* *Finding the nth Term* *Quadratic Sequences*		p. 12 ☐	p. 17 ☐	p. 27 ☐
G Perimeter and Area 1 *Perimeter and Area of Rectangles* *Area of a Triangle* *Area and Perimeter of Compound Shapes*		p. 18 ☐	p. 28 ☐	p. 38 ☐
G Perimeter and Area 2 *Area of a Parallelogram* *Area of a Trapezium* *Circumference and Area of a Circle*		p. 20 ☐	p. 28 ☐	p. 38 ☐
S Statistics and Data 1 *Mean, Median, Mode and Range* *Choosing which Average to Use* *Constructing a Tally Chart*		p. 22 ☐	p. 29 ☐	p. 39 ☐
S Statistics and Data 2 *Grouping Data* *Stem-and-Leaf Diagrams* *Two-Way Tables*		p. 24 ☐	p. 29 ☐	p. 39 ☐
N Decimals 1 *Multiplying and Dividing by 10, 100 and 1000* *Powers of Ten* *Ordering Decimals*		p. 30 ☐	p. 40 ☐	p. 50 ☐

N Number **A** Algebra **G** Geometry and Measures

	Topic	Revise	Practise	Review
N	Decimals 2 *Adding and Subtracting Decimals* *Multiplying Decimals* *Dividing Decimals* *Rounding and Estimating*	p. 32	p. 40	p. 50
A	Algebra 1 *Collecting Like Terms* *Expressions with Products* *Substitution*	p. 34	p. 41	p. 51
A	Algebra 2 *Expanding Brackets* *Factorising*	p. 36	p. 41	p. 51
G	3D Shapes: Volume and Surface Area 1 *Naming and Drawing 3D Shapes* *Using Nets to Construct 3D Shapes* *Surface Area of a Cuboid* *Volume of a Cuboid*	p. 42	p. 52	p. 62
G	3D Shapes: Volume and Surface Area 2 *Volume of a Cylinder* *Surface Area of a Cylinder* *Calculating the Volume of Composite Shapes*	p. 44	p. 52	p. 62
S	Interpreting Data 1 *Pie Charts* *Pictograms* *Frequency Diagrams* *Data Comparison*	p. 46	p. 53	p. 63
S	Interpreting Data 2 *Interpreting Graphs and Diagrams* *Drawing a Scatter Graph* *Statistical Investigations*	p. 48	p. 53	p. 63
N	Fractions 1 *Equivalent Fractions* *Adding and Subtracting Fractions*	p. 54	p. 64	p. 74
N	Fractions 2 *Multiplying and Dividing Fractions* *Mixed Numbers and Improper Fractions* *Adding and Subtracting Mixed Numbers*	p. 56	p. 64	p. 74

S Statistics P Probability R Ratio, Proportion and Rates of change

Contents

		Revise	Practise	Review
A	**Coordinates and Graphs 1** *Coordinates* *Linear Graphs* *Graphs of $y = ax + b$*	p. 58 ☐	p. 65 ☐	p. 75 ☐
A	**Coordinates and Graphs 2** *Gradients and Intercepts* *Solving Linear Equations from Graphs* *Drawing Quadratic Graphs*	p. 60 ☐	p. 65 ☐	p. 75 ☐
G	**Angles 1** *How to Measure and Draw an Angle* *Angles in a Triangle* *Angles in a Quadrilateral* *Bisecting an Angle*	p. 66 ☐	p. 76 ☐	p. 86 ☐
G	**Angles 2** *Angles in Parallel Lines* *Angles in Polygons* *Polygons and Tessellation*	p. 68 ☐	p. 76 ☐	p. 86 ☐
P	**Probability 1** *Probability Words* *Probability Scale* *Probability of an Event Not Occurring* *Sample Spaces*	p. 70 ☐	p. 77 ☐	p. 87 ☐
P	**Probability 2** *Mutually Exclusive Events* *Calculating Probabilities and Tabulating Events* *Experimental Probability* *Venn Diagrams and Set Notation*	p. 72 ☐	p. 77 ☐	p. 87 ☐
R	**Fractions, Decimals and Percentages 1** *Different Ways of Saying the Same Thing* *Converting Fractions to Decimals to Percentages* *Fractions of a Quantity* *Percentages of a Quantity*	p. 78 ☐	p. 88 ☐	p. 98 ☐
R	**Fractions, Decimals and Percentages 2** *Percentage Increase and Decrease* *Finding One Quantity as a Percentage of Another* *Simple Interest*	p. 80 ☐	p. 88 ☐	p. 98 ☐
A	**Equations 1** *Finding Unknown Numbers* *Solving Equations* *Equations with Unknowns on Both Sides*	p. 82 ☐	p. 89 ☐	p. 99 ☐
A	**Equations 2** *Solving More Complex Equations* *Setting Up and Solving Equations*	p. 84 ☐	p. 89 ☐	p. 99 ☐

Contents

			Revise	Practise	Review
G	**Symmetry and Enlargement 1**	*Reflection and Reflectional Symmetry* *Translation* *Rotational Symmetry* *Enlargement*	p. 90 ☐	p. 100 ☐	p. 110 ☐
G	**Symmetry and Enlargement 2**	*Congruence* *Scale Drawings* *Shape and Ratio*	p. 92 ☐	p. 100 ☐	p. 110 ☐
R	**Ratio and Proportion 1**	*Introduction to Ratios* *Simplifying Ratios*	p. 94 ☐	p. 101 ☐	p. 111 ☐
R	**Ratio and Proportion 2**	*Sharing Ratios* *Direct Proportion* *Using the Unitary Method*	p. 96 ☐	p. 101 ☐	p. 111 ☐
R	**Real-Life Graphs and Rates 1**	*Graphs from the Real World* *Reading a Conversion Graph* *Drawing a Conversion Graph*	p. 102 ☐	p. 112 ☐	p. 114 ☐
R	**Real-Life Graphs and Rates 2**	*Time Graphs* *Travelling at a Constant Speed* *Unit Pricing* *Density*	p. 104 ☐	p. 112 ☐	p. 114 ☐
G	**Right-Angled Triangles 1**	*Pythagoras' Theorem* *Finding the Longest Side* *Finding a Shorter Side*	p. 106 ☐	p. 113 ☐	p. 115 ☐
G	**Right-Angled Triangles 2**	*Side Ratios* *Finding Angles in Right-Angled Triangles* *Finding the Length of a Side*	p. 108 ☐	p. 113 ☐	p. 115 ☐
	Mixed Test-Style Questions			p. 116 ☐	
	Answers		p. 128		
	Glossary		p. 139		
	Index		p. 142

Number 1

You must be able to:

- Order positive and negative numbers
- Use the symbols $=, \neq, <, >, \leqslant, \geqslant$
- Multiply and divide integers
- Carry out operations following BIDMAS.

Positive and Negative Numbers

- A **number line** can be used to order **integers**.

Example
Place 7, 5, –6, –1 and 3 in ascending order.

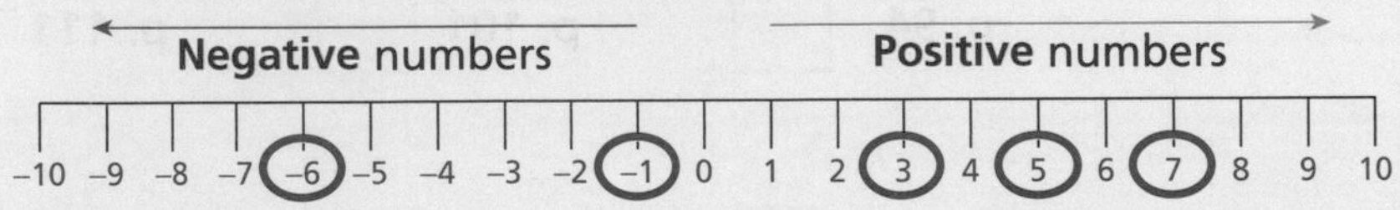

Going from left to right, you can see that the ascending order is –6, –1, 3, 5, 7.

Key Point

There is an infinite number of positive and negative numbers.

- **Place value** can be used to compare the size of large numbers.

Example
Which is greater, 3408 or 3540?

Number	Thousands	Hundreds	Tens	Units
3408	3	4	0	8
3540	3	5	4	0

Both numbers have 3 thousands, but 3540 has 5 hundreds and 3408 only has 4 hundreds. Therefore 3540 is greater than 3408.

Key Point

Always compare digits from left to right.

- **Symbols** are used to state the relationship between two numbers.

Symbol	Meaning
$>$	Greater than
$<$	Less than
$\geqslant$	Greater than or equal to
$\leqslant$	Less than or equal to
$=$	Equal to
$\neq$	Not equal to

Example
3540 is greater than 3408 can be written as $3540 > 3408$.

Multiplication and Division

- To multiply big numbers using the grid method, partition both numbers into their hundreds, tens and units.

Example
Calculate 354 × 273

×	300	50	4
200	60 000	10 000	800
70	21 000	3500	280
3	900	150	12

Then add together the numbers in the grid:

60 000 + 10 000 + 800 + 21 000 + 3500 + 280 + 900 + 150 + 12
= 96 642

So 354 × 273 = 96 642

354 is made up of 3 hundreds (300), 5 tens (50) and 4 units (4). 273 is made up of 2 hundreds (200), 7 tens (70) and 3 units (3).

Complete the multiplication grid, for example 50 × 70 = 3500 and 4 × 3 = 12

- Division can also be broken down into steps.

Example
Calculate 762 ÷ 3

$3\overline{)7\ 6\ 2}$	2 $3\overline{)7\ {}^{1}6\ 2}$	$2\ 5$ $3\overline{)7\ {}^{1}6\ {}^{1}2}$	$2\ 5\ 4$ $3\overline{)7\ {}^{1}6\ {}^{1}2}$
Set up your division.	Work out how many 3s go into 7. 3 goes into 7 twice with 1 left over. You carry the 1 over.	Work out how many 3s go into 16. 3 goes into 16 five times with 1 left over. You carry the 1 over.	Work out how many 3s go into 12. 3 goes into 12 four times with zero left over. **So 762 ÷ 3 = 254**

BIDMAS

- BIDMAS gives the order in which operations should be carried out:

Brackets
Indices
Divide
Multiply
Add
Subtract

Example
$2 \times 6 + 5 \times 4 = 32$

Key Point

Indices are also called powers.

Quick Test

1. Put 3750, 3753, 3601, 3654 and 3813 in ascending order.
2. Work out 435 × 521
3. Work out 652 ÷ 4
4. Work out 4 × 3 + 7 × 4
5. Write the following using a mathematical symbol: –5 is less than 3.

Key Words

integer
positive
negative
place value

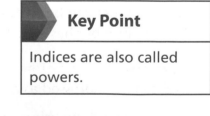

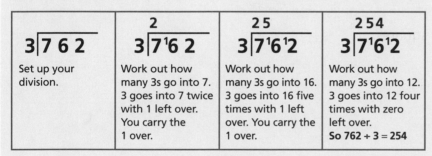

Number 2

You must be able to:

- Understand square numbers and square roots
- Write a number as a product of prime factors
- Find the lowest common multiple and highest common factor.

Squares and Square Roots

- **Square numbers** are calculated by multiplying a number by itself.

Example
$5^2 = 5 \times 5 = 25$

The first ten square numbers are:
1, 4, 9, 16, 25, 36, 49, 64, 81 and 100.

- A **square root** ($\sqrt{\ }$) is the inverse or opposite of a square.

Example
$\sqrt{36} = 6$

- There are other powers as well as squares.

Example
$5^3 = 5 \times 5 \times 5 = 125$

$5^4 = 5 \times 5 \times 5 \times 5 = 625$

$5^5 = 5 \times 5 \times 5 \times 5 \times 5 = 3125$

Prime Factors

- **Factors** are numbers you can multiply together to make another number.
- Every number can be written as a **product** of prime factors.
- A **prime** number has exactly two factors, itself and 1.

Example
45 can be expressed as a product of prime factors.

This can be done using a **prime factor tree**:

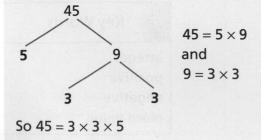

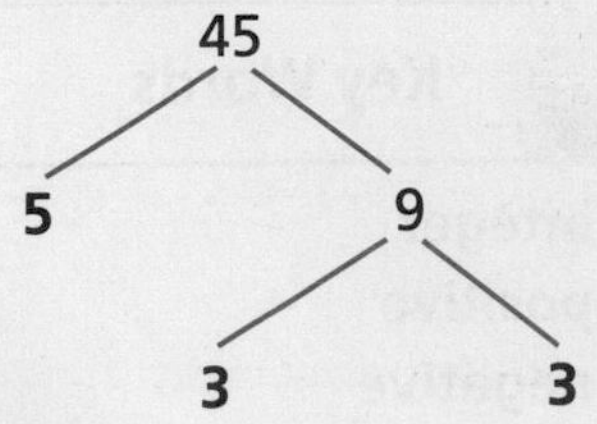

$45 = 5 \times 9$
and
$9 = 3 \times 3$

So $45 = 3 \times 3 \times 5$

Always start by finding a prime number which is a factor, in this case 5.

Remember to write the final answer as a product.

Key Point

Product means multiply.

Key Point

A prime factor tree breaks a number down into its prime factors.

Lowest Common Multiple and Highest Common Factor

- The **lowest common multiple** (LCM) is the lowest multiple two or more numbers have in common.
- The **highest common factor** (HCF) is the highest factor two or more numbers have in common.

Example
Find the lowest common multiple and highest common factor of 12 and 42.

Write both numbers as a product of prime factors.

$12 = 2 \times 2 \times 3$ $42 = 2 \times 3 \times 7$

Complete the Venn diagram.

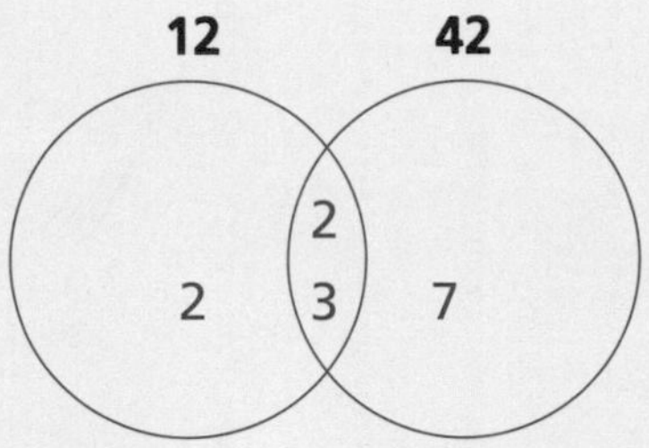

Common factors are placed in the overlap.

The LCM is the product of all the numbers in both circles.

LCM $= 2 \times 2 \times 3 \times 7 = 84$

The HCF is the product of the numbers in the overlap.

HCF $= 2 \times 3 = 6$

- LCM and HCF are used to solve many everyday problems.

Example
Thomas is training to swim the English Channel. He has to visit his doctor every 12 days and his nutritionist every 15 days. If on 1st October he has both appointments on the same day, on what date will he next have both appointments on the same day?

Find the LCM of 12 and 15 – this is 60.

60 days after 1st October is 30th November.

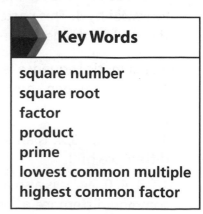

Quick Test

1. Write down the value of 7^2.
2. Write down the value of $\sqrt{49}$.
3. Write 40 as a product of prime factors.
4. Find the lowest common multiple of 14 and 36.
5. Find the highest common factor of 24 and 32.

Key Words

square number
square root
factor
product
prime
lowest common multiple
highest common factor

Sequences 1

You must be able to:

- Use a function machine to generate terms of a sequence
- Recognise arithmetic sequences
- Generate sequences from a term to term rule.

Function Machines

- A **function machine** takes an input, applies one or more operations and produces an output.

INPUT		OUTPUT
1		9
2	→ ×4 → +5	13
3		17
20		85

Sequences

- A **sequence** is a set of shapes or numbers which follow a pattern or rule.
- The outputs from a **function machine** form a sequence.

Example

In the sequence below, the next pattern is formed by adding an extra layer of tiles around the previous pattern.

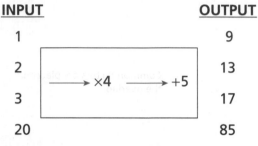

Basic design | Layer 1, 6 new tiles | Layer 2, 10 new tiles | Layer 3, 14 new tiles

With each new layer the number of new tiles needed increases by 4. This pattern can be used to predict how many tiles will be needed to make larger designs.

- An **arithmetic sequence** is a set of numbers with a common difference between consecutive terms.

Example

4, 7, 10, 13, 16,… is an arithmetic sequence with a common difference of 3.

12, 8, 4, 0, –4,… is also an arithmetic sequence. It has a common difference of –4.

Key Point

Arithmetic sequences are used to solve many real-life problems.

- There are many other sequences of numbers that follow a pattern.

Example
1, 4, 9, 16, 25, 36,… are square numbers.

1, 3, 6, 10, 15, 21,… are known as the triangular numbers.

1, 1, 2, 3, 5, 8, 13,… is known as the Fibonacci sequence. ← Each number is the sum of the previous two numbers.

Finding Missing Terms

- The **term to term rule** links each term in the sequence to the previous term.

Example
5, 8, 11, 14, 17,…

In this set of numbers the next term is found by adding 3 to the previous term. Therefore the term to term rule is +3.

This rule can be used to find the next numbers in the sequence.

$17 + 3 = 20$ $20 + 3 = 23$ $23 + 3 = 26$

Therefore the next three terms in the sequence are 20, 23, 26,…

- The term to term rule can also be used to find missing terms.

Example
13, 17, 21, ____, 29,…

The term to term rule is +4 and so the missing term is 25.

Quick Test

1. Find the missing inputs and outputs in the function machine.

   INPUT: 6, □ → ÷2 → +1 → OUTPUT: □, 25

2. Write down the next five terms in this sequence:
   5, 9, 13, 17,…
3. Write down the term to term rule.
   24, 12, 6, 3, 1.5,…
4. Which of the following sequences is arithmetic?
   A) 4, 6, 9, 13, 18,… **B)** 5, 9, 13, 17, 21,…
5. Find the missing term in the following sequence of numbers.
   21, 27, 33, ____, 45, 51,…

Key Words

function machine
sequence
arithmetic sequence
term to term

Sequences 2

You must be able to:
- Generate terms of a sequence from a position to term rule
- Find the nth term of an arithmetic sequence
- Recognise quadratic sequences.

The nth Term

- The nth term is also called the **position to term rule**.
- It is an algebraic expression that represents the operations carried out by a function machine.

INPUT		**OUTPUT**
1		9
2	→ × 4 → + 5	13
3		17
n		$4n + 5$

- The **nth term** can be used to generate terms of a sequence.

Example
The nth term of a sequence is given by $3n + 5$.

To find the first term you **substitute** $n = 1$

$3 \times 1 + 5 = 8$ 8 is the first term in the sequence.

To find other terms, you can substitute different values of n.

When $n = 2$	When $n = 3$	When $n = 4$
$3 \times 2 + 5 = 11$	$3 \times 3 + 5 = 14$	$3 \times 4 + 5 = 17$
So 11 is the second term in the sequence.	So 14 is the third term in the sequence.	So 17 is the fourth term in the sequence.

The nth term $3n + 5$ produces the sequence of numbers:

8, 11, 14, 17, 20,…

The rule can be used to find any term in the sequence. For example, to find the 50th term in the sequence substitute $n = 50$

$3 \times 50 + 5 = 155$

Key Point

For the first term in the sequence, n always equals 1.

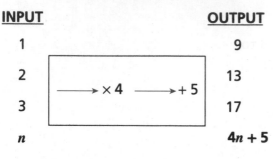

Finding the nth Term

- To find the nth term, look for a pattern in the sequence of numbers.

Example
The first five terms of a sequence are 7, 11, 15, 19, 23.

The term to term rule is +4 so the nth term starts with $4n$.

The difference between $4n$ and the output in each case is 3, so the final rule is $4n + 3$.

Input	$\times 4$	Output
1	4	7
2	8	11
3	12	15
4	16	19
5	20	23
n	$4n$	$4n + 3$

Quadratic Sequences

- Quadratic sequences are based on square numbers.

Example
The first five terms of the sequence $2n^2 + 1$ are as follows:

When $n = 1$ $2 \times (1)^2 + 1 = 3$
When $n = 2$ $2 \times (2)^2 + 1 = 9$
When $n = 3$ $2 \times (3)^2 + 1 = 19$
When $n = 4$ $2 \times (4)^2 + 1 = 33$
When $n = 5$ $2 \times (5)^2 + 1 = 51$

The sequence starts 3, 9, 19, 33, 51,…

- Triangular numbers are produced from a quadratic sequence.

Key Point

Use BIDMAS when calculating terms in a sequence.

Quick Test

1. Write down the first five terms in the sequence $5n + 3$.
2. Write down the first five terms in the sequence $n^2 + 4$.
3. a) Find the nth term for the following sequence of numbers: 6, 9, 12, 15, 18,…
   b) Find the 50th term in this sequence.
4. What is the nth term also known as?

Key Words

nth term
position to term
substitute
quadratic

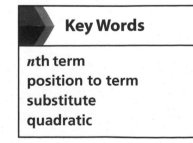

Review Questions

Key Stage 2: Key Concepts

1 Which is closer to 2000?

1996 or 2007 1996. The difference between 1996 and 2000 is 4. The difference between 2007 and 2000 is 7.

Explain how you know. [2]

2 Calculate 476 – 231 476 − 231 = 245 [2]

3 Complete the table below by rounding each number to the nearest 1000.

	To nearest 1000
4587	5000
45 698	46 000
457 658	458 000
45 669	46000

[2]

4 Write these in order starting with the smallest. [3]

0.56	55%	$\frac{27}{50}$	0.6	0.63
0.55	0.54			
(3)	(2)	(1)	(4)	(5)

5 Ahmed is twice as old as Rebecca.

Rebecca is three years younger than John.

John is 25 years old.

How old is Ahmed? 44 [2]

6 Calculate 467 × 34 [2]

(15878)

467
34
1868
14010
15878

7 Calculate 156 ÷ 3 [2]

3)156 = 52 (52)

8 On the scale below draw arrows to show 1.6 and 3.8

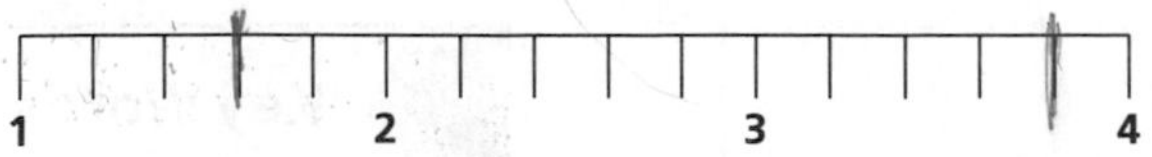

[2]

Total Marks .... / 17

1. A bottle holds 1 litre of fizzy drink. Mariam pours four glasses for her friends.

   Each glass contains 200ml.

   How much fizzy drink is left in the bottle? [3]

   200ml

2. Below are five digit cards.

7	5	1	6	3

   Choose two cards to make the following two-digit numbers.

   a) A square number 16 [1]

   b) A prime number 31 [1]

   c) A multiple of 6 36 [1]

   d) A factor of 60 15 [1]

3. An equilateral triangle has a perimeter of 27cm.

   What is the length of one of its sides? 9 [2]

4. Two-thirds of a number is 22.

   What is the number? 33 [2]

5. Here is an isosceles triangle drawn inside a rectangle.

   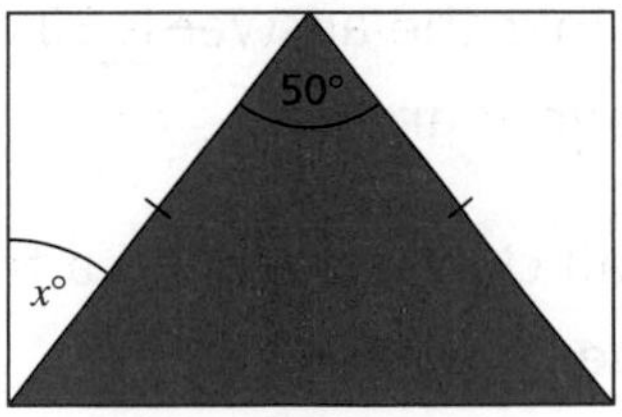

   Find the value of the angle x. [3]

   90 − 65 = 25

   25

6. S and T are two whole numbers.

   $S + T = 500$

   S is 100 greater than T.

   S = 300 T = 200

   Find the value of S and T. [2]

Total Marks .... / 16

Practice Questions

Number

1 The table shows the average daily minimum temperature and the average daily maximum temperature in a town for every month of the year (given to the nearest °C).

Month	Jan	Feb	Mar	Apr	May	Jun	Jul	Aug	Sep	Oct	Nov	Dec
Average Daily Minimum Temp.	–8	–4	–1	3	8	11	14	13	9	3	–2	–6
Average Daily Maximum Temp.	–1	4	11	17	21	25	28	27	22	13	5	0

a) In which month is the average daily minimum temperature lowest? [1]

Jan

b) In which month is the average daily maximum temperature highest? [1]

Jul

c) In March, what is the difference between the average daily minimum temperature and the average daily maximum temperature? 12 [1]

PS 2 The area of a square is 49cm^2. Work out the length of a side. [1]

7

Total Marks .... / 4

MR 1 Jessa and Holly have been given the following question:

BIDMAS

What is the value of $3 + 5 \times 4 + 7$? 30

Jessa thinks the answer is 30 and Holly thinks the answer is 39. Who is right? Explain your answer. Jessa because she used BIDMAS [2]

FS 2 A netball club is planning a trip. The club has 354 members and the cost of the trip is £12 per member.

a) Work out the total cost of the trip. [3]

4248

They need coaches for the trip and each coach seats 52 people.

b) How many coaches do they need to book? [3]

7

c) How many spare seats will there be? [2]

10

MR 3 Explain why $\sqrt{79}$ must be between 8 and 9. [2]

It is between $\sqrt{64}$ and $\sqrt{81}$

Total Marks .... / 12

Sequences

1 **a)** Write down the next two numbers in the following sequence.

3, 7, 11, 15, 19, 23,... [2]

b) Write down the term to term rule. [1]

4n - 1

2 Lynne plants a new flower in her garden.

Of the buds, five have already flowered.

Each week another three buds flower.

a) How many buds will have flowered after three weeks? [1]

b) How many weeks will it take for 32 buds to have flowered? [1]

Total Marks .... / 5

1 **a)** Find the nth term of this arithmetic sequence.

4, 7, 10, 13, 16,... [3]

3n + 1

b) Find the 60th term in the sequence. [1]

181

MR **2** Match the cards on the left with a card on the right.

5, 9, 13, 17, 21...	Neither
2, 8, 18, 32, 50...	Quadratic
8, 17, 32, 53, 80...	Arithmetic

[2]

Total Marks .... / 6

Geometry and Measures

Perimeter and Area 1

You must be able to:

- Find the perimeter and area of a rectangle
- Find the area of a triangle
- Find the area and perimeter of compound shapes.

Perimeter and Area of Rectangles

- The **perimeter** is the distance around the outside of a 2D shape.
- The formula for the perimeter of a rectangle is:

  perimeter = 2(length + width) or $P = 2(l + w)$

  also perimeter = 2(length) + 2(width)

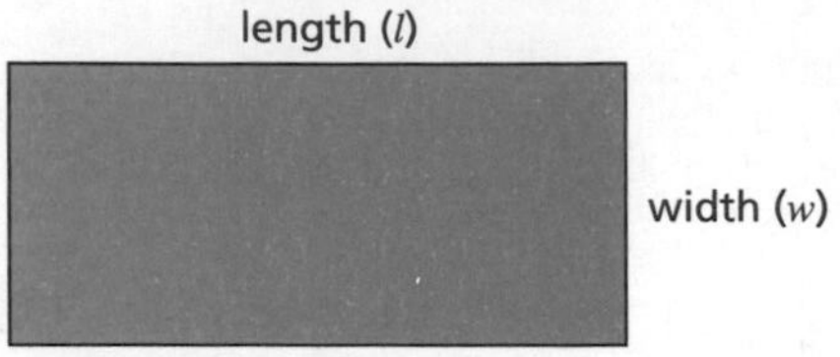

- The formula for the **area** of a rectangle is:

  area = length × width or $A = l \times w$

Example

Find the perimeter and area of this rectangle.

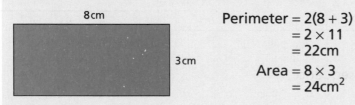

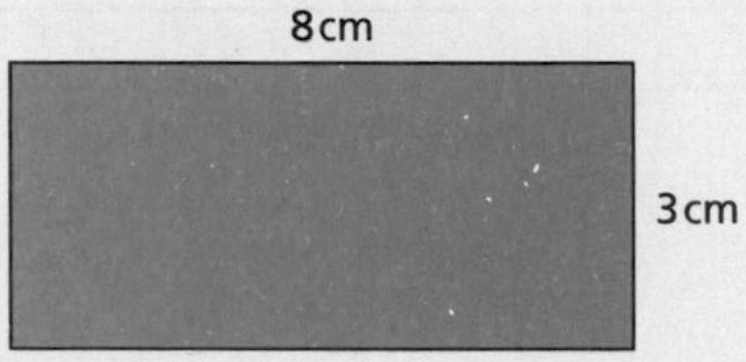

Perimeter = 2(8 + 3)
= 2 × 11
= 22cm

Area = 8 × 3
= 24cm^2

Area of a Triangle

- The formula for the area of a triangle is:

  area = $\frac{1}{2}$(base × **perpendicular** height)

Example

Find the area of this triangle.

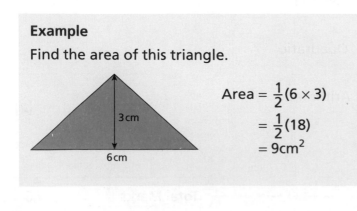

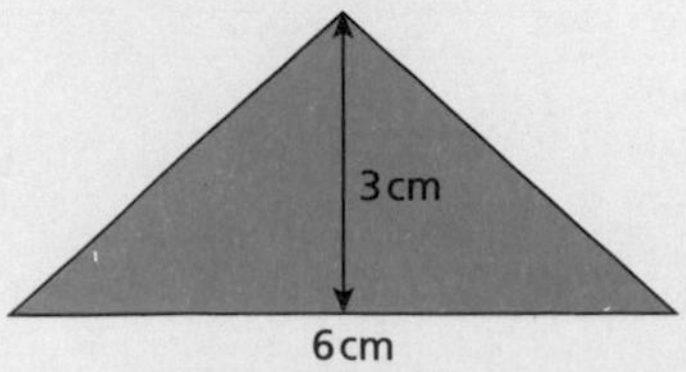

$$\text{Area} = \frac{1}{2}(6 \times 3) = \frac{1}{2}(18) = 9\text{cm}^2$$

Key Point

When finding the area of a triangle, always use the perpendicular height.

Area and Perimeter of Compound Shapes

- A **compound** shape is made up from other, simpler shapes.
- To find the area of a compound shape, divide it into basic shapes.

Example

This shape can be broken up into three rectangles.

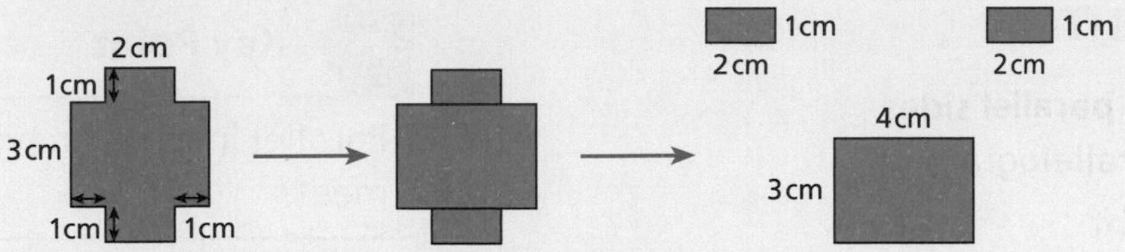

The areas of the individual rectangles are $2cm^2$, $2cm^2$ and $12cm^2$.

The area of the compound shape is $2 + 2 + 12 = 16cm^2$.

Key Point

Areas are two-dimensional and are measured in square units, for example cm^2.

- To find the perimeter, start at one corner of the shape and travel around the outside, adding the lengths.

Example

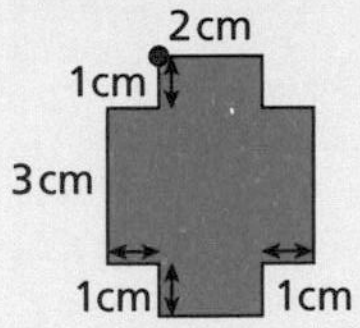

Perimeter

$= 2 + 1 + 1 + 3 + 1 + 1 + 2 + 1 + 1 + 3 + 1 + 1$

$= 18cm$

Quick Test

1. Find the perimeter of a rectangle with width 5cm and length 7cm.
2. Find the area of a rectangle with width 9cm and length 3cm. Give appropriate units in your answer.
3. Find the area of a triangle with base 4cm and perpendicular height 3cm.
4. Find the perimeter and area of this shape.

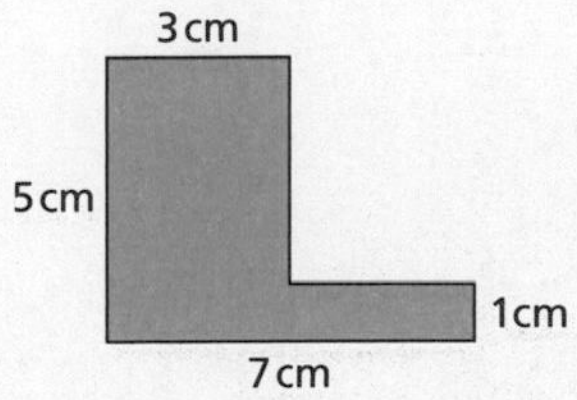

Key Words

perimeter
area
perpendicular
compound

Geometry and Measures

Perimeter and Area 2

You must be able to:

- Find the area of a parallelogram
- Find the area of a trapezium
- Find the circumference and area of a circle.

Area of a Parallelogram

- A **parallelogram** has two pairs of **parallel** sides.
- The formula for the area of a parallelogram is:

  area = base × perpendicular height

Key Point

Parallel lines will never meet.

Example

Find the area of this parallelogram.

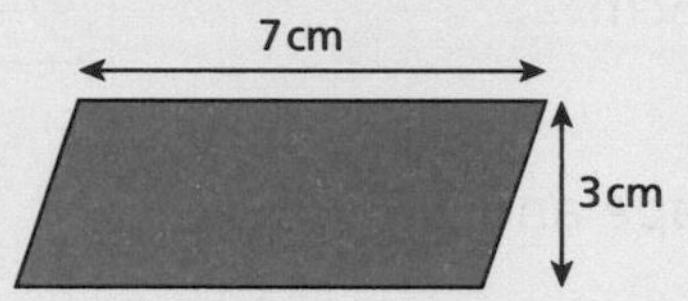

The base is 7cm and the perpendicular height is 3cm.

Area = 7 × 3

= 21cm^2

Area of a Trapezium

- A **trapezium** has one pair of parallel sides.
- The formula for the area of a trapezium is:

  $A = \frac{1}{2}(a + b)h$
- The sides labelled ***a*** and ***b*** are the parallel sides and ***h*** is the perpendicular height.
- Perpendicular means at right angles.

Example

Find the area of this trapezium.

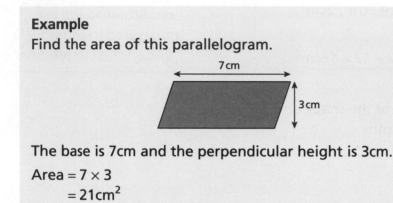

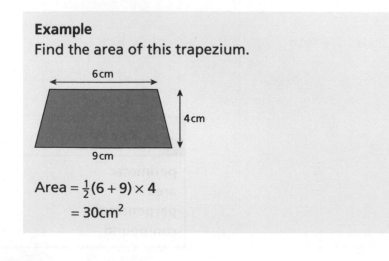

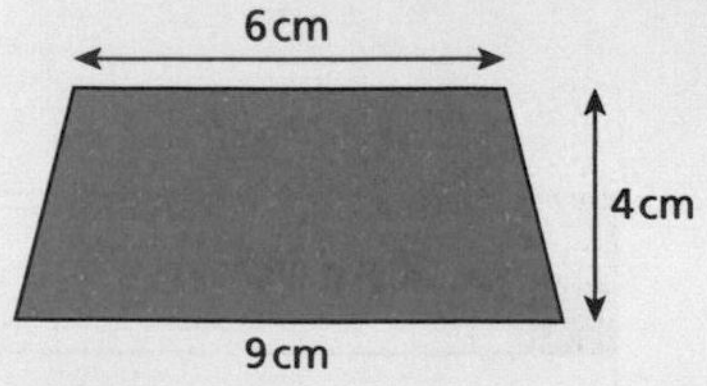

Area = $\frac{1}{2}(6 + 9) \times 4$

= 30cm^2

Circumference and Area of a Circle

- The **circumference** of a circle is the distance around the outside.
- The radius of a circle is the distance from the centre to the circumference.
- The diameter of a circle is twice the radius.
- The formula for the circumference of a circle is:
  $C = 2\pi r$ or $C = \pi d$
- The formula for the area of a circle is:
  $A = \pi r^2$

Example
Find the circumference and area of this circle.
Give your answers to 1 decimal place.

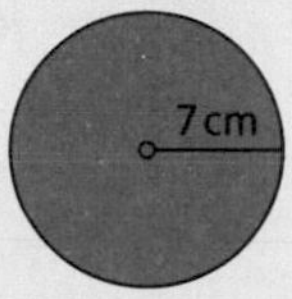

The circumference:
$C = 2 \times \pi \times 7$
$= 14 \times \pi$
$= 44.0\text{cm}$

The area:
$A = \pi \times 7^2$
$= \pi \times 49$
$= 153.9\text{cm}^2$

- Circles can be split into **sectors**.
- A sector is a region bounded by two radii and an arc (a curved line that is part of the circumference).
- A sector with a 90° angle at the centre would have an area of $\frac{1}{4}$ of the whole circle.

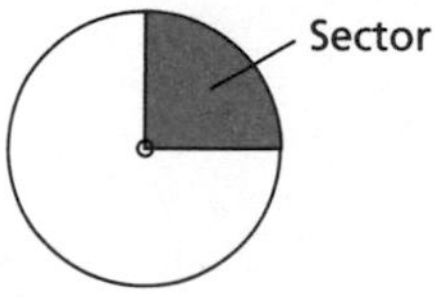

Key Point

The symbol π represents the number pi (3.141 592 654...).

π is approximately 3.14 or $\frac{22}{7}$

Quick Test

1. Find the area of the parallelogram.

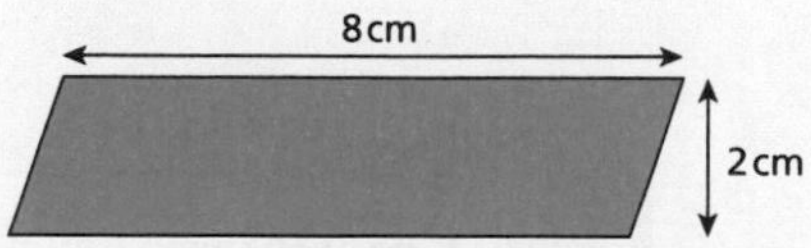

2. Find the area of the trapezium.

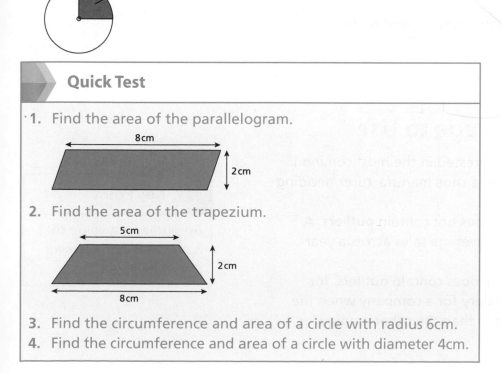

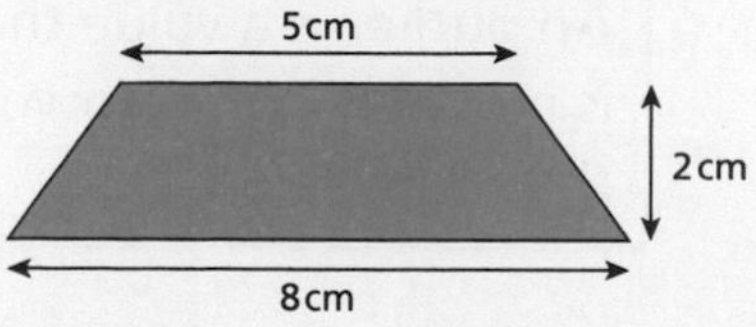

3. Find the circumference and area of a circle with radius 6cm.
4. Find the circumference and area of a circle with diameter 4cm.

Key Words

parallelogram
parallel
trapezium
pi (π)
circumference
sector

Statistics

Statistics and Data 1

You must be able to:

- Find the mean, median, mode and range for a set of data
- Choose which average is the most appropriate to use in different situations
- Use a tally chart to collect data.

Mean, Median, Mode and Range

- The **mean** is the **sum** of all the values divided by the number of values.
- The **median** is the middle value when the data is in order.
- The **mode** is the most common value.
- The mean, median and mode are all averages.
- The **range** is the **difference** between the biggest and the smallest value.
- The range is a measure of spread.

Key Point

Data can have more than one mode. Bi-modal means the data set has two modes.

Example

Find the mean, median, mode and range for the following set of data:

5, 9, 7, 6, 2, 7, 3, 11, 4, 2, 6, 6, 4

The mean $= \dfrac{5+9+7+6+2+7+3+11+4+2+6+6+4}{13} = 5.5...$

Mean = 5.5 (1 d.p.)

The median = 2, 2, 3, 4, 4, 5, ⑥ 6, 6, 7, 7, 9, 11

Put the data in order, smallest to biggest.

Median = 6

The mode is also 6 as this number is seen most often.

The range = 11 − 2 = 9

Choosing which Average to Use

- Use the **mode** when you are interested in the most common answer, for example if you were a shoe manufacturer deciding how many of each size to make.
- Use the **mean** when your data does not contain **outliers**. A company which wanted to find average sales across a year would want to use all values.
- Use the **median** when your data does contain outliers, for example finding the average salary for a company when the manager earns many times more than the other employees.

Key Point

An outlier is a value that is much higher or lower than the others.

Constructing a Tally Chart

- A tally **chart** is a quick way of recording data.
- Your data is placed into groups, which makes it easier to analyse.
- A tally chart can be used to make a **frequency** chart by adding an extra column to record the total in each group.

Key Point

The frequency is the total for the group.

Example
You are collecting and recording data about people's favourite flavour crisps. You ask 50 people and fill in the tally chart with their responses.

Flavour	Tally	Frequency
Plain	𝍸 𝍸 ||	12
Salt and vinegar	𝍸 ||||	9
Cheese and onion	𝍸 𝍸 𝍸 |	16
Prawn cocktail	𝍸 ||	7
Other	𝍸 |	6
Total		**50**

From the table it can be seen that cheese and onion is the mode.

Quick Test

1. Emma surveyed her class to find out their favourite colour. She constructed the tally chart below.
   a) Complete the frequency column.
   b) How many pupils are there in Emma's class?

Colour	Tally	Frequency
Red	𝍸 ||||	
Blue	𝍸 ||	
Green	𝍸 |||	
Yellow	𝍸 |	
Other	𝍸 ||	

2. Look at this set of data:
   3, 7, 4, 6, 3, 5, 9, 40, 6
   a) Write down the mode of this data.
   b) Calculate the mean, median and range.
   c) Would you choose the mean or the median to represent this data? Explain your answer.

Key Words

mean
sum
median
mode
range
difference
outlier
chart
frequency

Statistics

Statistics and Data 2

You must be able to:

- Group data and construct grouped frequency tables
- Draw a stem-and-leaf diagram
- Construct and interpret a two-way table.

Grouping Data

- When you have a large amount of data it is sometimes appropriate to place it into groups.
- A group is also called a **class interval**.
- The disadvantage of using **grouped data** is that the original **raw data** is lost.

Example
The data below represents the number of people who visited the library each day over a 60-day period.

Number of people	Frequency
0–10	10
11–20	30
21–30	14
Over 30	6

On 30 out of the 60 days, the library had between 11 and 20 (inclusive) visitors.

- Data such as the number of people is discrete as it can only take particular values.
- Data such as height and weight is continuous as it can take any value on a particular scale.

Key Point

Calculations based on grouped data will be estimates.

Stem-and-Leaf Diagrams

- A stem-and-leaf diagram is a way of organising the data without losing the **raw data**.
- The data is split into two parts, for example tens and units.
- In the example below, the stem is the tens and the leaves the units.
- Each row should be ordered from smallest to biggest.
- A stem-and-leaf diagram must have a **key**.

	KEY: 1 | 2 = 12
0	1 6 6 7 9 — This is the number 9
1	2 2 4 5 8 8 9
2	3 2 5 5 7
3	0 2 — This is the number 32

Key Point

A stem-and-leaf diagram orders data from smallest to biggest.

Two-Way Tables

- A two-way table shows information that relates to two different categories.
- Two-way tables can be constructed from information collected in a survey.

Example

Daniel surveyed his class to find out if they owned any pets. In his class there are 16 boys and 18 girls. 10 of the boys owned a pet and 15 of the girls owned a pet.

	Pets	No Pets	Total
Boys	10		16
Girls	15		18
Total			

The information given is filled into the table and then the missing information can be worked out.

If there are 16 boys in Daniel's class and 10 of them have pets, then to work out how many boys do not have pets we calculate $16 - 10 = 6$

When all the missing information is entered you can **interpret** the data.

	Pets	No Pets	Total
Boys	10	6	16
Girls	15	3	18
Total	25	9	34

You can see that there are 34 pupils in Daniel's class and 25 of them owned a pet. You can also see that more girls owned pets than boys, and that more pupils owned pets than did not own pets.

Key Point

Two-way tables can also be used to estimate probabilities.

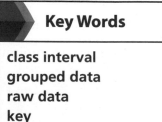

Quick Test

1. 25 women and 30 men were asked if they preferred football or rugby. 16 of the women said they prefer football and 10 of the men said they prefer rugby.
   a) Construct a two-way table to represent this information.
   b) How many in total said they prefer football?
   c) How many women preferred rugby?
   d) How many people took part in the survey?

Key Words

class interval
grouped data
raw data
key
interpret

Review Questions

Number

(FS) **1** I purchase a new car for £3000.

I pay a deposit of £800 and pay the rest in four equal payments.

How much is each payment? [3]

2 The planets in our solar system orbit the Sun.
Their orbits are almost circular.
The table gives the time it takes, in days, for each planet to complete one orbit of the Sun.

Earth	365
Jupiter	4332
Mars	687
Mercury	88
Neptune	60 200
Saturn	10 760
Uranus	30 700
Venus	224

Put the planets in order, starting with the shortest orbit time. [2]

3 A square has a perimeter of 20cm.

Work out the area of the square. [2]

4 Work out $4 \times 2^2 + 6$ [2]

Total Marks .... / 9

(MR) **1** $48 \times 52 = 2496$
Use this to help you work out the following calculations.

$24 \times 52 =$ .... $48 \times$ .... $= 1248$ $2496 \div 52 =$ .... [3]

(PS) **2** The lowest common multiple of two numbers is 60 and their sum is 27.

What are the numbers? [2]

(FS) **3** Gautam is having a barbecue and wants to invite some friends.

Sausages come in packs of 6. Rolls come in packs of 8.

She needs exactly the same number of sausages and rolls.

What is the minimum number of each pack she can buy? [3]

Total Marks .... / 8

Sequences

1 a) A function machine maps the number n to $n + 4$.

Fill in the missing values.

n	→	$n + 4$
5	→	
	→	20

[2]

b) Many different function machines can map 36 to 6. Complete the boxes below to give two **different** function machines.

n	→		n	→	
36	→	6	36	→	6

[2]

Total Marks .... / 4

1 An expression for the nth term of the arithmetic sequence 6, 8, 10, 12, … is $2n + 4$.

a) Find the 20th term of this sequence. [1]

b) Find the 100th term of this sequence. [1]

c) The odd numbers also form an arithmetic sequence with a common difference of 2. Find the nth term for the sequence of odd numbers. [2]

PS **2** Lundy Island has two lighthouses; one in the north and one in the south.

The northern lighthouse flashes every 20 seconds and the southern lighthouse every 45 seconds.

Both lighthouses flash at 4pm.

When do they next both flash at the same time? [3]

Total Marks .... / 7

Practice Questions

Perimeter and Area

1 Work out the perimeter and area of the rectangle below.

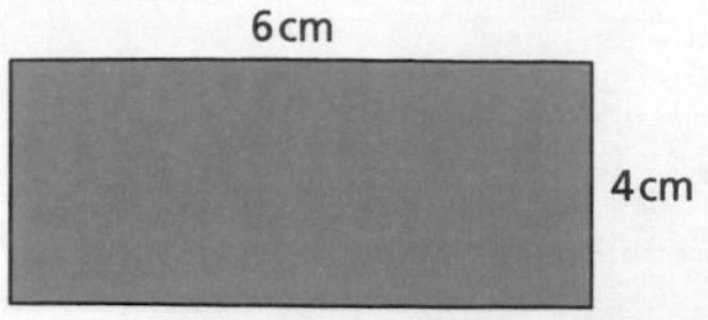

[2]

2

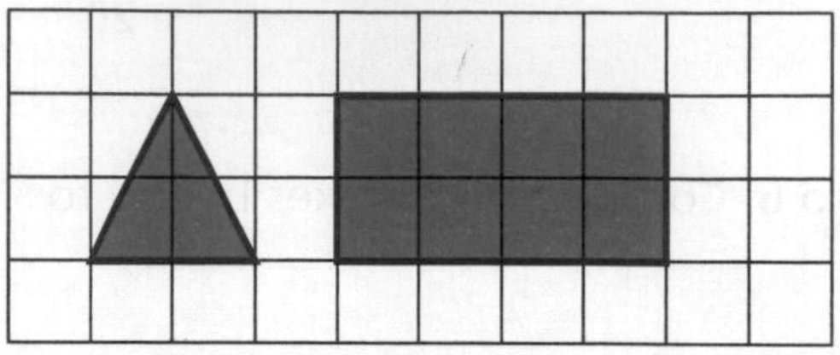

In the diagram above the area of the triangle is $8cm^2$.

Use this to help you work out the area of the rectangle. [2]

Total Marks .... / 4

PS 1 The area of this rectangle is $48cm^2$.

Find the value of X and Y.

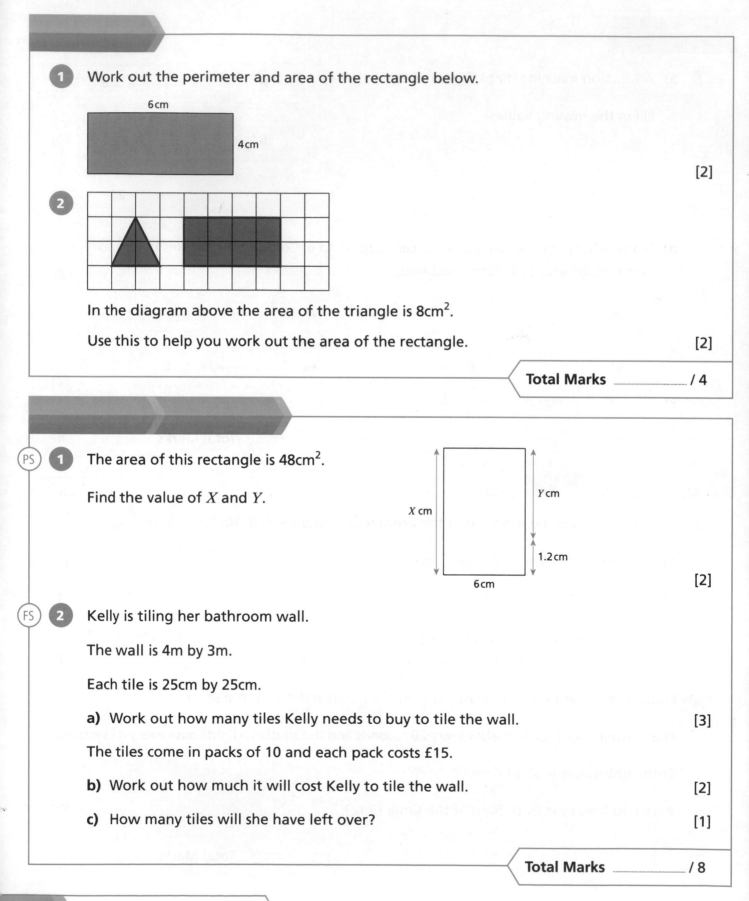

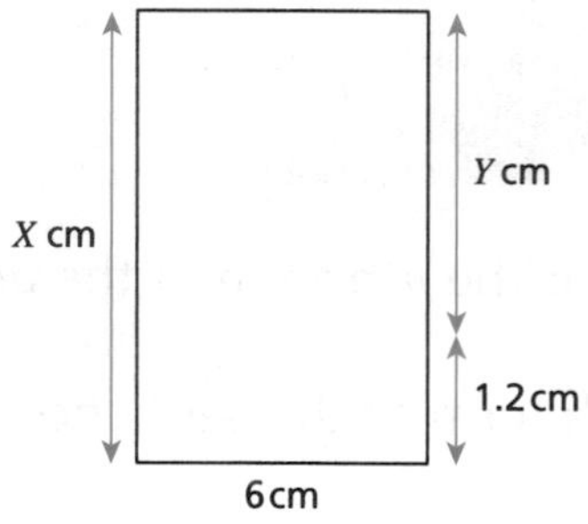

[2]

FS 2 Kelly is tiling her bathroom wall.

The wall is 4m by 3m.

Each tile is 25cm by 25cm.

a) Work out how many tiles Kelly needs to buy to tile the wall. [3]

The tiles come in packs of 10 and each pack costs £15.

b) Work out how much it will cost Kelly to tile the wall. [2]

c) How many tiles will she have left over? [1]

Total Marks .... / 8

Statistics and Data

1 Look at these five numbers: 6 11 9 12 7

a) Show that the mean of the five numbers is 9. [1]

b) Explain why the median is also 9. [1]

c) Write down a different set of five numbers which also have a mean of 9. [1]

PS 2 The scores for a netball team are 15, 23, 8, 5, 19, x.

The mean score is 15. What is x? [2]

3 This stem-and-leaf diagram shows the marks of some students in a test.

4	1 5 6 7 8 9
5	2 3 4 6 8 9 9 9
6	3 4 5 5 8 9
7	0 2 6

KEY: 5 | 2 = 52 marks

A student was absent on the day the test was taken, took it at a later date and scored 78.

a) Add this score to the stem-and-leaf diagram. [1]

b) Write down the mode. [1]

c) Work out the range of the scores. [1]

Total Marks .... / 8

MR 1 Phil and Dave are both good darts players.

Their scores for a match are shown below.

Phil	64	70	80	100	57	100	41	56	30
Dave	36	180	21	180	10	5	23	25	140

a) Calculate the mean score for each player. [2]

b) Find the range of scores for each player. [2]

c) Only one of the two players can be picked to play in the next match.

Would you pick Phil or Dave? Explain your answer. [2]

Total Marks .... / 6

Decimals 1

You must be able to:

- Multiply and divide by 10, 100 and 1000
- Understand the powers of 10
- List decimals in size order.

Multiplying and Dividing by 10, 100 and 1000

- Multiplying by 10 moves the **decimal point** one place to the right.
- Multiplying by 100 moves the decimal point two places to the right.
- Multiplying by 1000 moves the decimal point three places to the right.

Example

$1.67 \times 10 = 16.7$ $1.67 \times 100 = 167$ $1.67 \times 1000 = 1670$

Key Point

Multiplying by 10, 100 or 1000 makes the number bigger.

- Dividing by 10 moves the decimal point one place to the left.
- Dividing by 100 moves the decimal point two places to the left.
- Dividing by 1000 moves the decimal point three places to the left.

Example

$360.75 \div 10 = 36.075$ $360.75 \div 100 = 3.6075$

$360.75 \div 1000 = 0.36075$

Key Point

Dividing by 10, 100 or 1000 makes the number smaller.

Powers of Ten

- A **power** or **index** tells us how many times a number should be multiplied by itself.

Example

$10^2 = 10 \times 10 = 100$

$10^3 = 10 \times 10 \times 10 = 1000$

$10^4 = 10 \times 10 \times 10 \times 10 = 10\,000$

$10^{-1} = \frac{1}{10}$ $10^{-2} = \frac{1}{10^2} = \frac{1}{100}$

- **Standard form** allows us to write very big and very small numbers more easily. Standard form uses powers of 10.
- A number not written in standard form is an **ordinary number**.
- 2000 can be written as 2×1000, which is the same as 2×10^3.
- When writing in standard form, the first digit must be 1 or more but less than 10.

Ordering Decimals

- Place value can be used to compare decimal numbers.
- The numbers after the decimal point are called tenths, hundredths, thousandths, and so on.

Example

Put these numbers in order from smallest to biggest:

12.071, 12.24, 12.905, 12.902, 12.061

Each number starts with 12. So compare the tenths, hundredths and thousandths.

	Tens	Units	.	Tenths	Hundredths	Thousandths
12.071	1	2	.	0	7	1
12.24	1	2	.	2	4	0
12.905	1	2	.	9	0	5
12.902	1	2	.	9	0	2
12.061	1	2	.	0	6	1

First group by the number of tenths.

12.071, 12.061 are the two smallest as they have no tenths.

12.24 is the next smallest with 2 tenths.

12.905 and 12.902 are the two biggest as they have 9 tenths.

Then order them within each group.

12.061 is smaller than 12.071 as it has only 6 hundredths compared to 7 hundredths.

12.902 is smaller than 12.905, as although they both have the same hundredths, 12.902 has only 2 thousandths compared to 5 thousandths.

So from smallest to biggest:

12.061, 12.071, 12.24, 12.902, 12.905

Key Point

Ascending order is smallest to biggest.

Descending order is biggest to smallest.

Quick Test

1. Work out 23.56 × 10
2. Work out 56.781 ÷ 10
3. Write down the value of 10^5.
4. Write these numbers in ascending order:
   16.34, 16.713, 16.705, 16.309, 16.2

Key Words

decimal point
power
index
standard form
ordinary number

Decimals 2

You must be able to:

- Add and subtract decimal numbers
- Multiply and divide decimal numbers
- Use rounding to estimate calculations.

Adding and Subtracting Decimals

- Decimal numbers can be added and subtracted in the same way as whole numbers.

Example
Calculate 23.764 + 12.987

	2	3	.	7	6	4
+	1	2	.	9	8	7
	3	6	.	7	5	1
		1		1	1	

So 23.764 + 12.987 = 36.751

Example
Calculate 12.697 – 8.2

	~~1~~	¹2	.	6	9	7
–	0	8	.	2	0	0
		4	.	4	9	7

So 12.697 – 8.2 = 4.497

Key Point

When adding and subtracting, line up the numbers by matching the decimal point.

Multiplying Decimals

- Complete the calculation without the decimal points and replace the decimal point at the end.
- Count how many numbers are after the decimal points in the question and this is how many numbers are after the decimal point in the answer.

Example
Calculate 45.3 × 3.7

453 × 37 = 16 761

There are two numbers after the decimal point in the question.

So 45.3 × 3.7 = 167.61

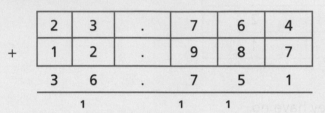

Dividing Decimals

- **Equivalent** fractions can be used when dividing decimals.

Example

Calculate 4.45 ÷ 0.05

$4.45 \div 0.05 = \frac{4.45}{0.05} = \frac{445}{5}$

$$\begin{array}{r} 8\ 9 \\ 5\overline{)4^{4}4^{4}5} \end{array}$$

So 4.45 ÷ 0.05 = 89

Remember to multiply the numerator and denominator by the same amount.

Key Point

A division can be written as a fraction. Equivalent fractions are equal.

Rounding and Estimating

- Numbers can be **rounded** using **decimal places** (d.p.) or **significant figures** (s.f.).

Example

Round 56.76 to 1 decimal place.

56.<u>7</u>|6

7 is the first decimal place and the number after it is more than 5 so round 7 up to 8. 56.76 to 1 decimal place is 56.8

- When **estimating** a calculation, round all the numbers to 1 s.f.

Example

Estimate 26 751 × 64

Round <u>2</u>|6 751

2 is the first significant figure. As the number after it is 5 or more, round the 2 up to 3 and every number after becomes zero.

So 26 751 to 1 significant figure is 30 000.

Round <u>6</u>|4

6 is the first significant figure. As the number after it is less than 5, the 6 does not change and every number after becomes zero.

So 64 to 1 significant figure is 60.

An estimate for 26 751 × 64 is 30 000 × 60 = 1 800 000

Key Point

The first significant figure is the first non-zero digit.

- There is always an error to consider when a number is rounded.
- This error can be expressed using an inequality.
- If a number is rounded to 23 to the nearest whole number, the actual number could be anywhere between 22.5 and 23.5. The rounding error can be expressed as $-0.5 \leqslant \text{error} < 0.5$.

Key Words

equivalent
rounding
decimal places
significant figures
estimate

Quick Test

1. Work out 45.671 + 3.82
2. Work out 34.321 – 17.11
3. Work out 65.2 ÷ 0.4
4. Estimate 3457 × 46

Algebra 1

You must be able to:

- Know the difference between an equation and expression
- Collect like terms in an expression
- Write products as algebraic expressions
- Substitute numerical values into formulae and expressions.

Collecting Like Terms

- The difference between an **equation** and an **expression** is that an equation has an equals sign.
- To **simplify** an expression like terms are collected.

Example
Simplify $2x + 6y - x + 4y$

Collect the like terms:

$2x - x + 6y + 4y$

The x terms can be simplified: $2x - x = x$

The y terms can be simplified: $6y + 4y = 10y$

So $2x + 6y - x + 4y$ can be simplified to $x + 10y$

Example
Simplify $2x^2 + 6y - x^2 + 4y - 6$

Collect the like terms:

$2x^2 - x^2 + 6y + 4y - 6$

The x terms can be simplified: $2x^2 - x^2 = x^2$

The y terms can be simplified: $6y + 4y = 10y$

There is only one constant term.

So $2x^2 + 6y - x^2 + 4y - 6$ can be simplified to $x^2 + 10y - 6$

Example
Simplify $\frac{2}{3}x + y - \frac{1}{3}x + \frac{3}{4}y$

Collect the like terms:

$\frac{2}{3}x - \frac{1}{3}x + y + \frac{3}{4}y$

The x terms can be simplified: $\frac{2}{3}x - \frac{1}{3}x = \frac{1}{3}x$

The y terms can be simplified: $y + \frac{3}{4}y = \frac{7}{4}y$

So $\frac{2}{3}x + y - \frac{1}{3}x + \frac{3}{4}y$ can be simplified to $\frac{1}{3}x + \frac{7}{4}y$

Key Point

Remember that terms have a + or – sign between them and each sign belongs to the term on its right.

Expressions with Products

- **Product** means multiply.
- Expressions with products are written without the × sign.

Example

$2 \times a = 2a$

$a \times b = ab$

$a \times a \times b = a^2b$

$a \div b = \frac{a}{b}$

$a \times a \times a = a^3$

Substitution

- A **formula** is a rule which links a variable to one or more other variables.
- The variables are written in shorthand by representing them with a letter.
- Some commonly used scientific formulae are:

speed = $\frac{\text{distance}}{\text{time}}$ in shorthand $s = \frac{d}{t}$

density = $\frac{\text{mass}}{\text{volume}}$ in shorthand $d = \frac{m}{v}$

- Substitution involves replacing the letters in a given formula or expression with numbers.

Example

Find the value of the expression $2a + b$ when $a = 3$ and $b = 5$.

Replace the letters with the given numbers.

$2a + b = 2 \times 3 + 5 = 11$

Example

Emma's mum lives 30 miles from her and on a particular morning last week Emma's journey there took $\frac{3}{4}$ of an hour.

Calculate her average speed in miles per hour.

$s = \frac{d}{t}$

$s = \frac{30}{0.75} = 40\text{mph}$

$\frac{3}{4} = 0.75$

Key Point

There are many scientific problems which involve substituting into formulae. Always follow BIDMAS.

To find the area and volume of shapes, we substitute into a formula.

Quick Test

1. Simplify $4x + 7y + 3x - 2y + 6$
2. Simplify $c \times c \times d \times d$
3. Find the value of $4x + 2y$ when $x = 2$ and $y = 3$

Key Words

equation
expression
simplify
product
formula

Algebra

Algebra 2

You must be able to:

- Multiply a single term over a bracket
- Factorise linear expressions.

Expanding Brackets

- **Expanding** the brackets involves removing the brackets by **multiplying** every term inside the bracket by the number or term on the outside.
- This table is a reminder of the rules when multiplying and dividing negative numbers.

	+	−
+	+	−
−	−	+

Example

Expand $3(x+5)$

×	3
x	$3x$
$+5$	$+15$

$3(x+5) = 3x+15$

Example

Expand and simplify $4(x+y)-2(2x-3y)$

×	4
x	$4x$
y	$4y$

$= 4x + 4y$

×	2
$2x$	$4x$
$-3y$	$-6y$

$= 4x - 6y$

Then collect like terms.

$(4x+4y)-(4x-6y) = 10y$

Key Point

Make sure you multiply every term inside the bracket.

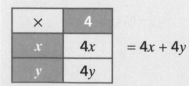

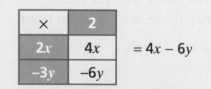

Factorising

- **Factorising** is the opposite of expanding.
- When factorising we put the brackets back in.

Example

Factorise $6x + 9$

3 is a common factor of 6 and 9 so we take the 3 to the outside of the bracket.

To find what is inside the bracket we need to fill in the blanks in the table.

×	3
	$6x$
	$+9$

×	3
$2x$	$6x$
$+3$	$+9$

So $6x + 9 = 3(2x + 3)$

Example

Factorise $x^2 + 2x$

x is the common factor:

×	x
	x^2
	$+2x$

×	x
x	x^2
$+2$	$+2x$

So $x^2 + 2x = x(x + 2)$

Example

Factorise $6x^3 + 2x^2$

2 and x^2 are the common factors:

×	$2x^2$
	$6x^3$
	$2x^2$

×	$2x^2$
$3x$	$6x^3$
$+1$	$2x^2$

So $6x^3 + 2x^2 = 2x^2(3x + 1)$

Key Point

Always expand your answer to check you are right.

Quick Test

1. Expand $4(2x - 1)$
2. Expand and simplify $2(2x - y) - 2(x + 6y)$
3. Factorise $5x - 25$
4. Factorise completely $2x^2 - 4x$

Key Words

expand
factorise

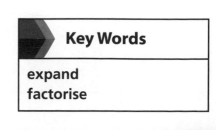

Review Questions

Perimeter and Area

1 Find the area of a circle with radius 7cm. [2]

2 Find the perimeter and area of the rectangle below.

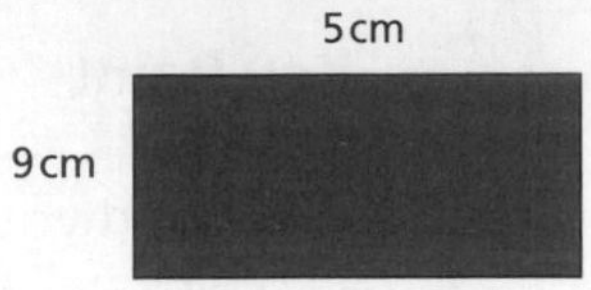

[2]

3 Find the area of the trapezium below.

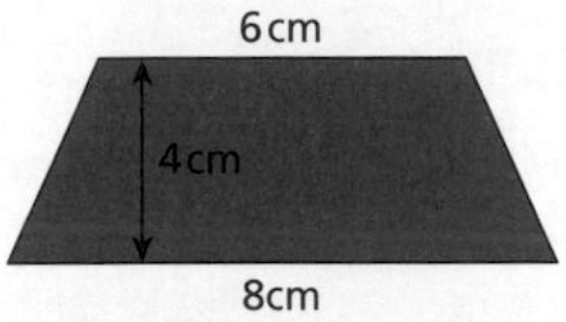

[2]

Total Marks .... / 6

1 Frances wants to paint the front of her house.
The diagram represents the front of her house.

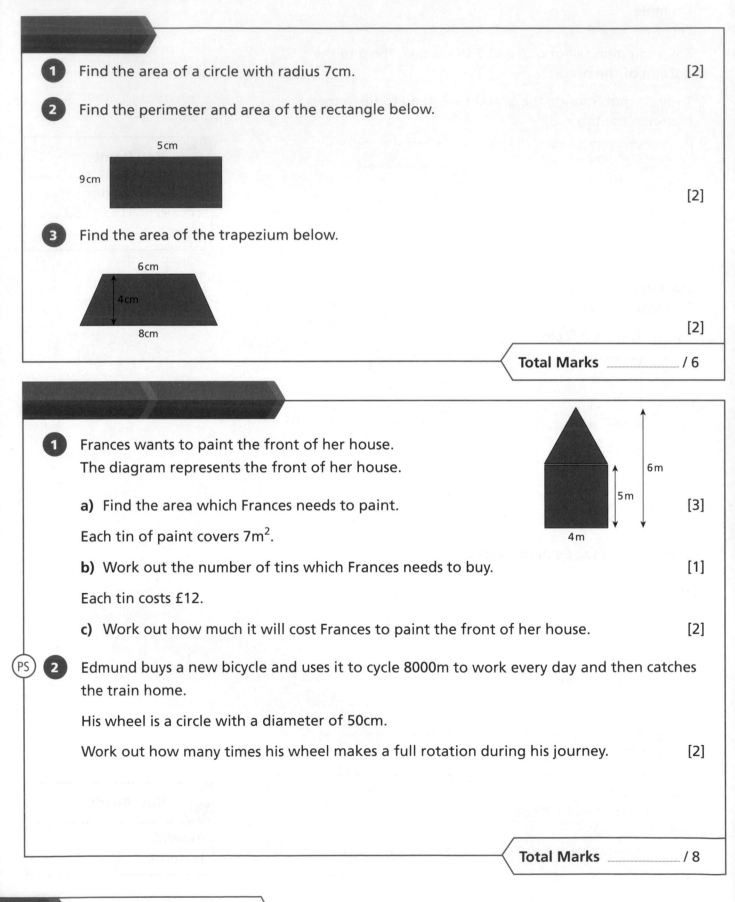

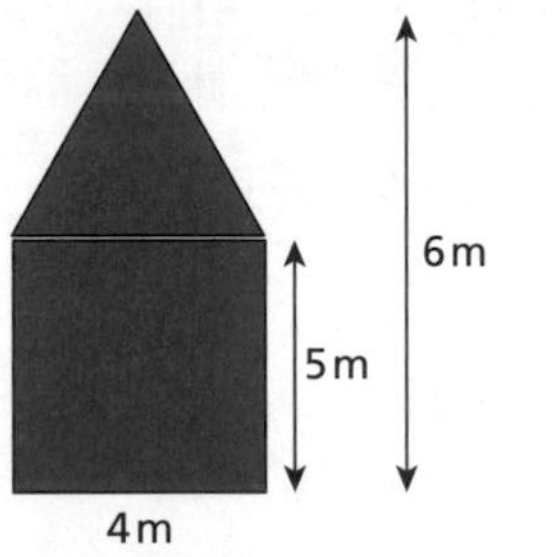

a) Find the area which Frances needs to paint. [3]

Each tin of paint covers $7m^2$.

b) Work out the number of tins which Frances needs to buy. [1]

Each tin costs £12.

c) Work out how much it will cost Frances to paint the front of her house. [2]

PS **2** Edmund buys a new bicycle and uses it to cycle 8000m to work every day and then catches the train home.

His wheel is a circle with a diameter of 50cm.

Work out how many times his wheel makes a full rotation during his journey. [2]

Total Marks .... / 8

Statistics and Data

1 There are 34 pupils in class 8D and 19 of those are boys.

26 of the class are right-handed and 14 of those are girls.

Use this information to complete the table below. [3]

	Boys	Girls
Right-handed		
Left-handed		

2 James does a survey of the favourite vegetables of the pupils in his class:

carrots	*potatoes*	*peas*	*sweetcorn*	*peas*	*peas*
potatoes	*peas*	*carrots*	*peas*	*potatoes*	*sweetcorn*
potatoes	*potatoes*	*potatoes*	*carrots*	*carrots*	*potatoes*
peas	*potatoes*	*sweetcorn*	*peas*	*potatoes*	*peas*

a) Construct a frequency table to represent this data. [2]

b) What was the mode? [1]

Total Marks .... / 6

MR 1 This data below shows the number of people attending six matches for Sandex United.

1240 **1354** **1306** **14 808** **1378** **1430**

You want to calculate the average attendance.

a) Would you find the mean, median or mode? Give a reason for your answer. [2]

b) Which of the attendance numbers is an outlier? [1]

Total Marks .... / 3

Practice Questions

Decimals

1 Match the card on the left with the correct card on the right.

One has been done for you.

3.7×10		0.037
3.7×100	→	370
$3.7 \div 10$		37
$3.7 \div 100$		0.37

[2]

2 Which is greater, 0.5679 or 0.5588?

Explain your answer. [2]

Total Marks .... / 4

1 Work out the following:

a) $34.542 + 23.29$ [1]

b) $65.21 - 43.23$ [1]

c) 21.81×3.4 [1]

d) $43.2 \div 0.2$ [1]

Total Marks .... / 4

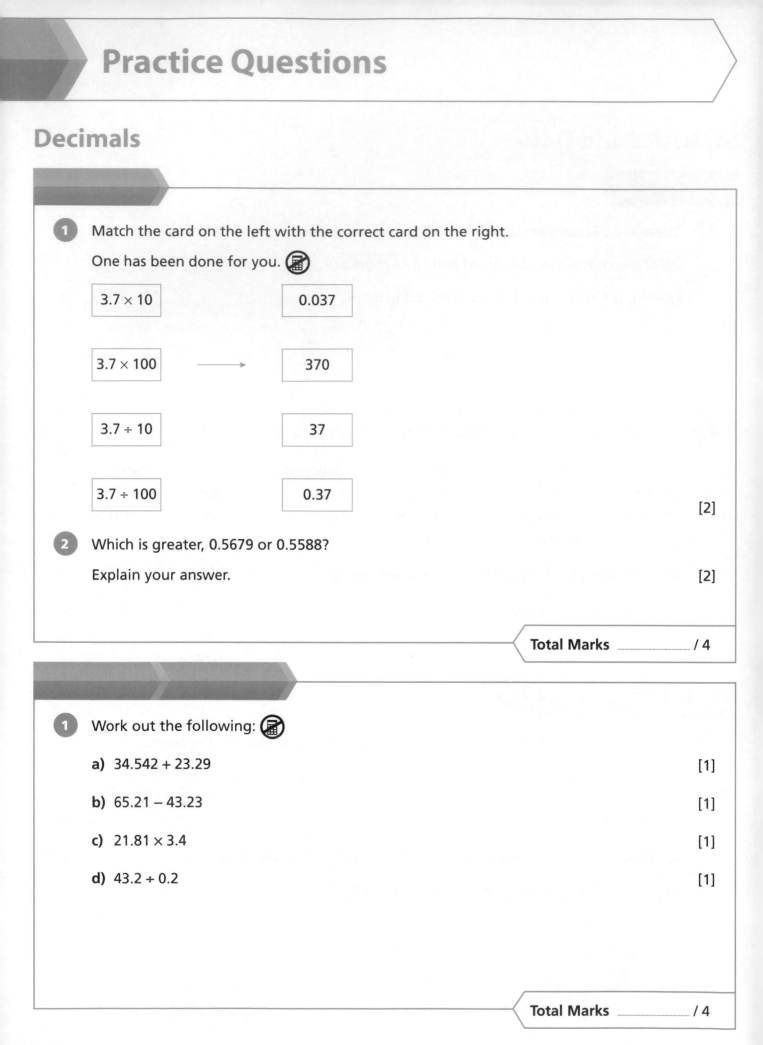

Algebra

1 Complete the statements.

When $x = 5$ $5x =$ ....

When $x =$ .... $5x = 45$

When $x = 7$ .... $= 28$ [2]

2 In this algebra grid, the total on each brick is made from the sum of the two bricks below. The first brick has been calculated for you. Complete the grid.

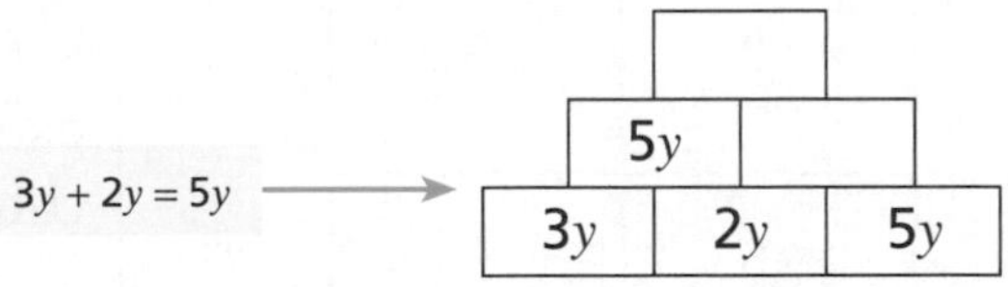

[2]

Total Marks .... / 4

MR 1 Claudia thinks the perimeter of this rectangle is $2x + 2y$.
Lawrence thinks the perimeter of the rectangle is $2(x + y)$.

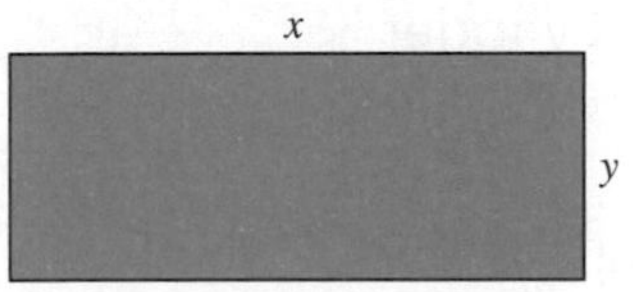

Who is right – Claudia, Lawrence or both of them?
Explain your answer. [2]

FS 2 The cost (in £) of hiring a car for the day and driving it y miles is shown by this formula:

$C = 75 + 0.4y$

Work out how much it would cost to hire the car for a day and travel 120 miles. [3]

3 Expand and simplify $2(4x + 1) - 5(x - 1)$. [2]

4 Factorise completely $3abc + 6a$. [2]

PS 5 If $ab = 36$ and $a = 4$, find the value of a^2b. [2]

6 Complete the algebra grid below. The total in each brick is made from the sum of the two bricks below.

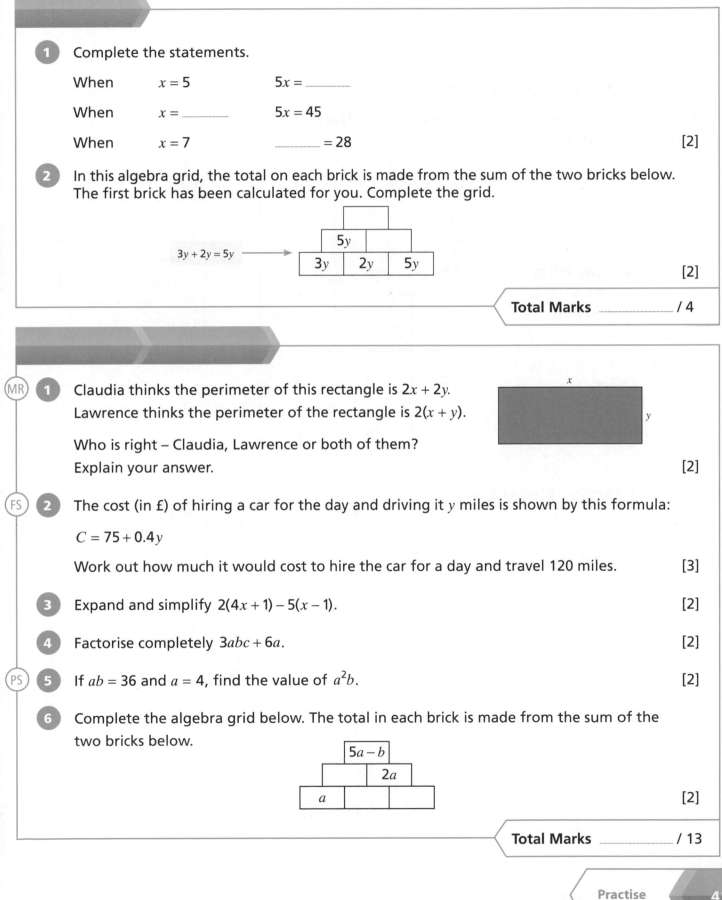

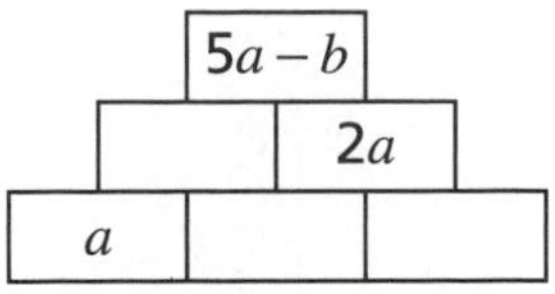

[2]

Total Marks .... / 13

3D Shapes: Volume and Surface Area 1

You must be able to:

- Name and draw 3D shapes
- Draw the net of a cuboid and other 3D shapes
- Calculate the surface area and volume of a cuboid.

Naming and Drawing 3D Shapes

- A 3D shape can be described using the number of **faces, vertices** and **edges** it has.

Shape	Name	Edges	Vertices	Faces
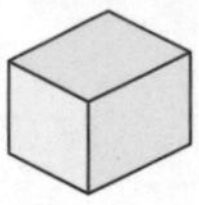	Cuboid	12	8	6
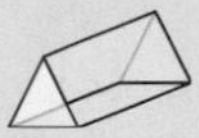	Triangular **prism**	9	6	5
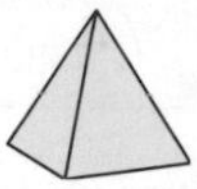	Square-based pyramid	8	5	5
	Cylinder	2	0	3
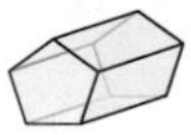	Pentagonal prism	15	10	7

Key Point

A face is a sur'face', for example a flat side of a cube.

A vertex is where edges meet, for example the corner of a cube.

An edge is where two faces join.

Using Nets to Construct 3D Shapes

- To create the **net** of a cuboid, imagine it is a box you are unfolding to lay out flat.

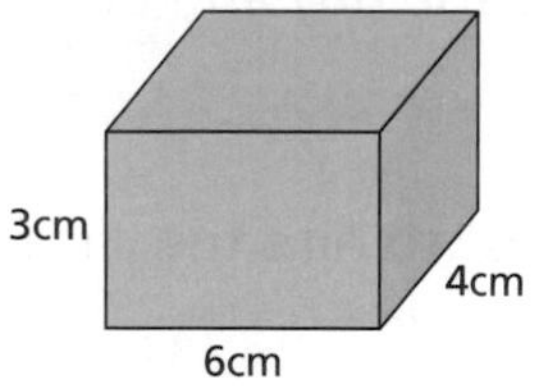

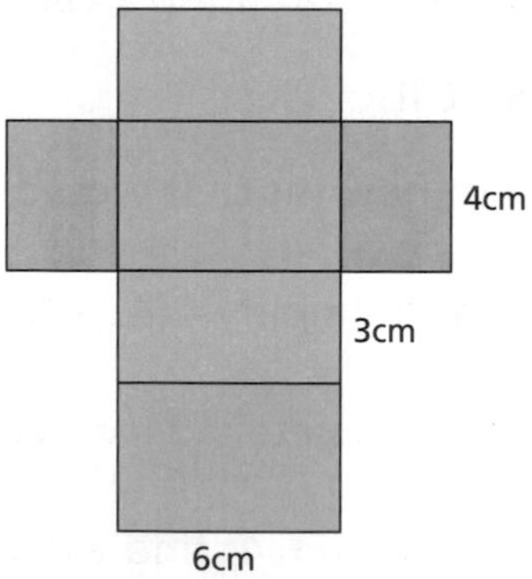

- Nets for a triangular prism and a square-based pyramid look like this:

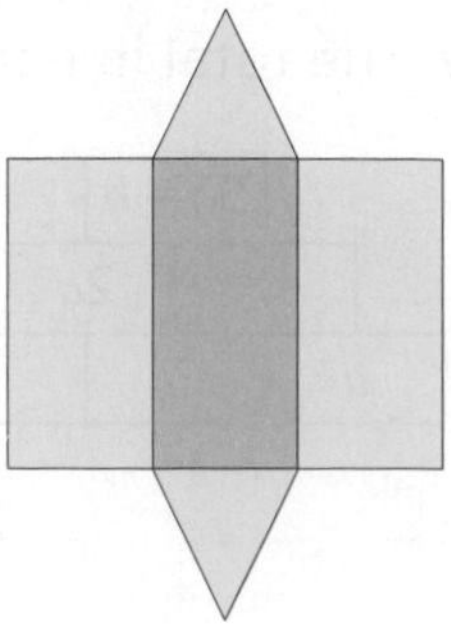

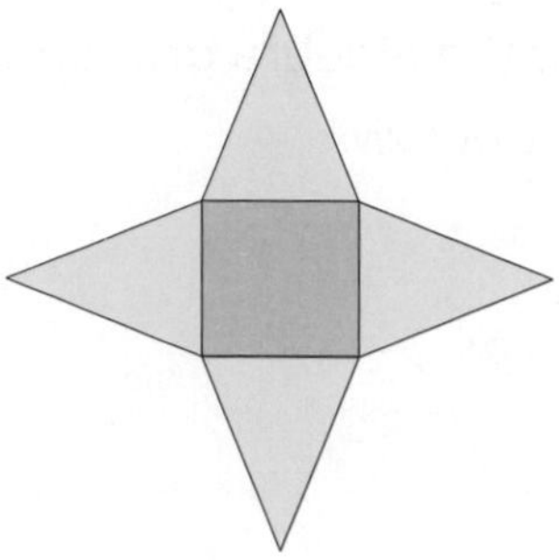

Surface Area of a Cuboid

- The **surface area** of a cuboid is the sum of the areas of all six faces and is measured in square units (cm^2, m^2, etc.).

> **Example**
> To calculate the surface area of the cuboid on the previous page, you can use the net to help you.
>
> Work out the area of each rectangle by multiplying the base by its height:
>
> 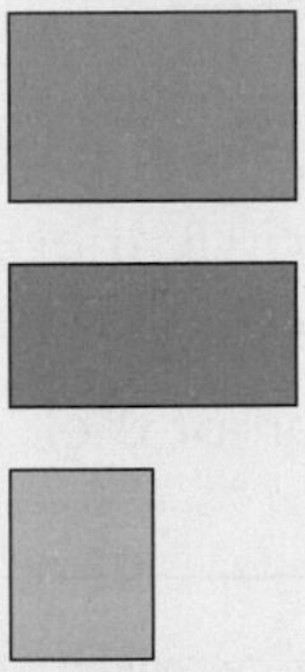
>
> Green rectangle: $6cm \times 4cm = 24cm^2$
> There are two of them, so $24cm^2 \times 2 = 48cm^2$
>
> Blue rectangle: $3cm \times 6cm = 18cm^2$
> There are two of them, so $18cm^2 \times 2 = 36cm^2$
>
> Pink rectangle: $4cm \times 3cm = 12cm^2$
> There are two of them, so $12cm^2 \times 2 = 24cm^2$
>
> Sum of all six areas is $48 + 36 + 24 = 108cm^2$

Volume of a Cuboid

- **Volume** is the space contained inside a 3D shape.

> **Example**
> To calculate the volume of the cuboid on the previous page, you need to first work out the area of the front rectangle:
>
> $3cm \times 6cm = 18cm^2$
>
> Next multiply this area by the depth of the cuboid, 4cm.
>
> $18cm^2 \times 4cm = 72cm^3$

The units are cm^3 this time.

Quick Test

Work out the volume and the surface area of these cuboids.

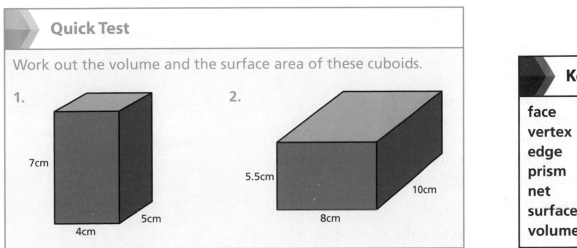

1.

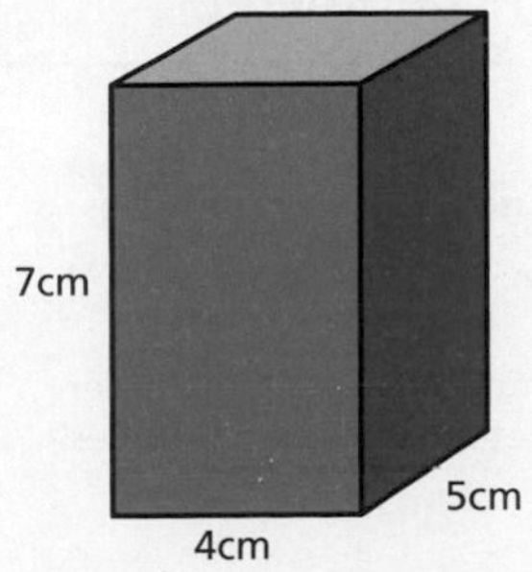

2.

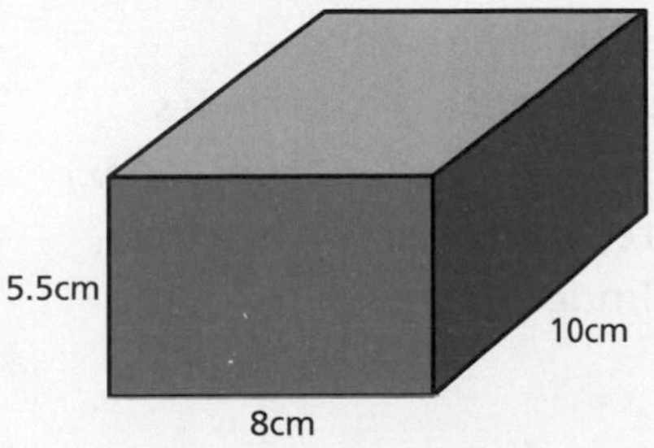

Key Words

face
vertex
edge
prism
net
surface area
volume

3D Shapes: Volume and Surface Area 2

You must be able to:

- Calculate the volume of a cylinder
- Calculate the surface area of a cylinder
- Calculate the volume of composite shapes.

Volume of a Cylinder

- You work out the volume of a **cylinder** the same way as the volume of a cuboid. First work out the area of the **circle** and then multiply it by the height of the cylinder.

Example

Calculate the volume of this cylinder.

Volume = $(\pi \times 5 \times 5) \times 3 = 235.62\text{cm}^3$ (2 d.p.)

This cylinder has a **diameter** of 12cm.

This means the **radius** is 6cm.

Volume $(\pi \times 6 \times 6) \times 5 = 565.49\text{cm}^3$ (2 d.p.)

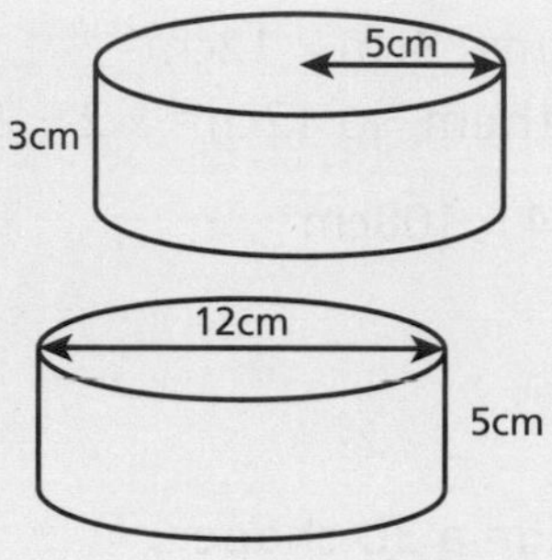

Key Point

Diameter is the full width of a circle that goes through the centre.

Radius is half of the diameter.

Area of a circle = $\pi \times$ radius²

Volume units are shown by a ^{3}, for example cm^3.

Surface Area of a Cylinder

- To calculate the surface area, first draw the net of a cylinder (imagine cutting a can open).

Example

Calculate the surface area of this cylinder.

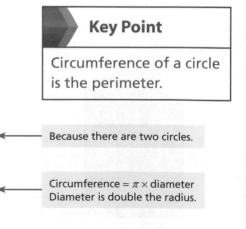

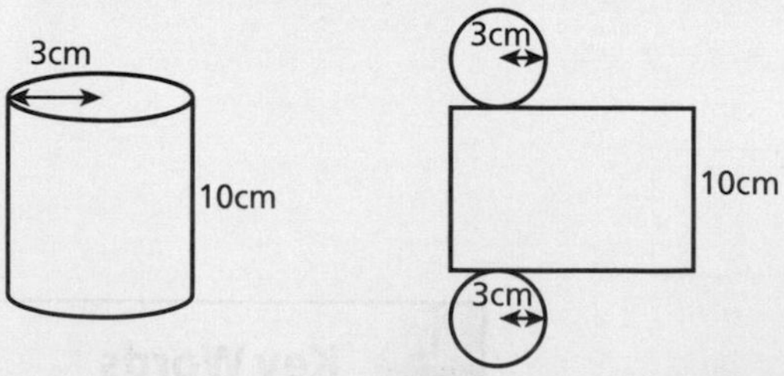

The net is made up from two circles and a rectangle.
You need to find the area and circumference of the circles.
Calculate the area of the circular ends $(\pi \times 3 \times 3) \times 2 = 56.55$ (2 d.p.)
The area of the rectangle is the circumference of the circle multiplied by the height of the cylinder, 10cm.

$(\pi \times 6) \times 10 = 188.50$ (2 d.p.)

Then add the areas together: $56.55 + 188.50 = 245.05\text{cm}^2$

Key Point

Circumference of a circle is the perimeter.

Because there are two circles.

Circumference = $\pi \times$ diameter
Diameter is double the radius.

Calculating the Volume of Composite Shapes

- **Composite** means the shape has been 'built' from more than one shape.

Example

This shape is built from two cuboids.

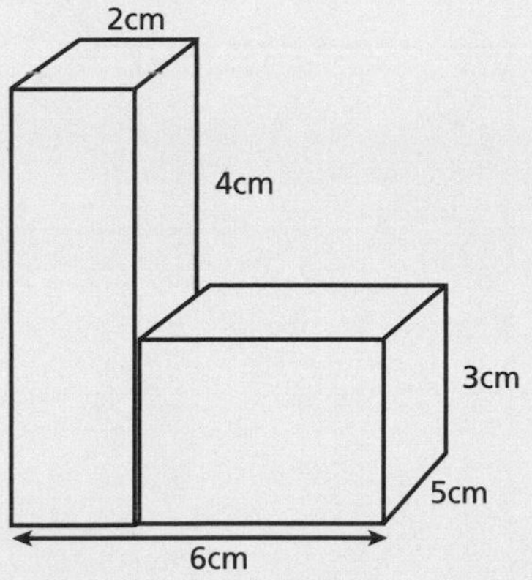

Calculate the volume of the two separate cuboids and add the volumes together.

$(6\text{cm} - 2\text{cm}) \times 5 \times 3 = 60\text{cm}^3$
$(3\text{cm} + 4\text{cm}) \times 2 \times 5 = 70\text{cm}^3$
$60 + 70 = 130\text{cm}^3$

Key Point

Use the shape's dimensions to work out missing lengths.

Quick Test

1. Work out the volume and the surface area of these shapes.

a)

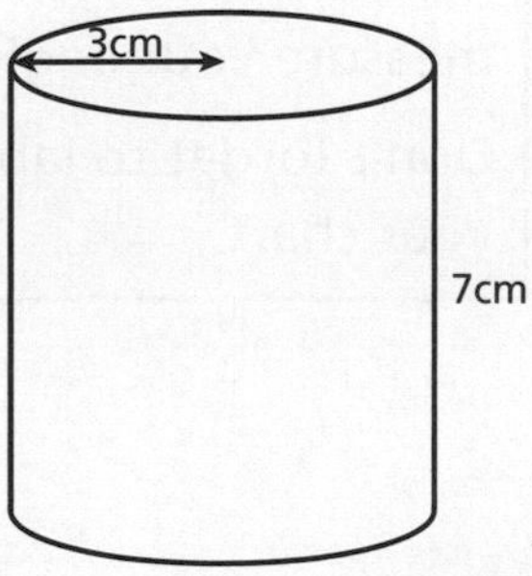

b)

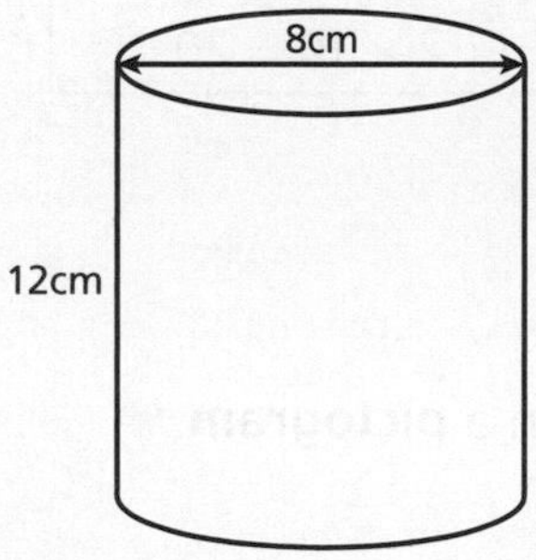

2. Work out the volume of this composite shape.

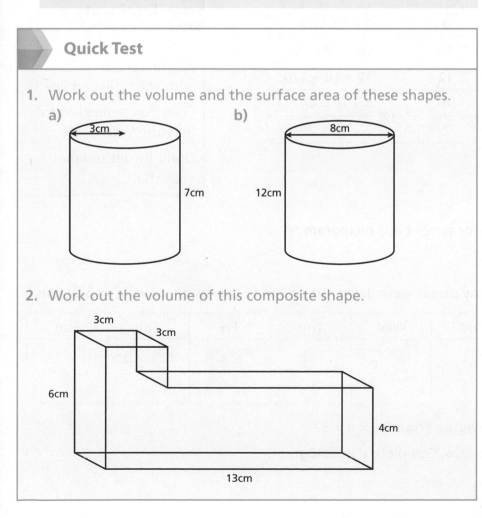

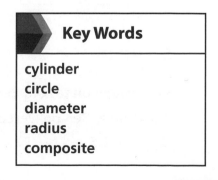

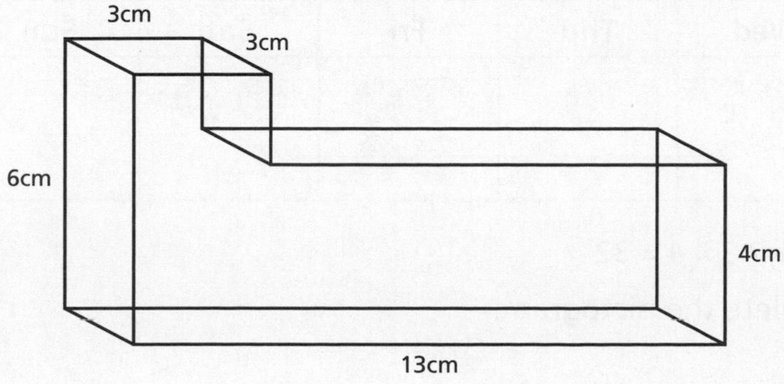

Key Words

cylinder
circle
diameter
radius
composite

Interpreting Data 1

You must be able to:

- Create a simple pie chart from a set of data
- Create and interpret pictograms
- Draw a frequency diagram
- Make comparisons and contrasts between data.

Pie Charts

- **Pie charts** are often shown with **percentages** or **angles** indicating sector size – this and the visual representation helps to **interpret** the **data**.

Example

36 students were asked the following question:

Which is your favourite flavour of crisps?

To work out the angle for each sector: 360 ÷ 36 = 10°

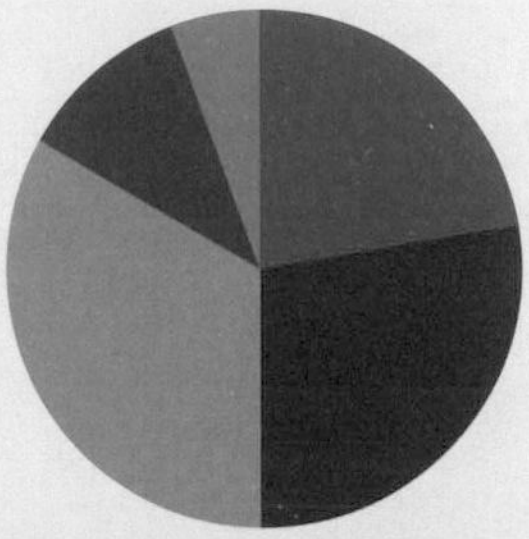

$$\frac{\text{Degrees in full turn}}{\text{Total}} = \text{Degrees per person}$$

Flavour of crisps	No. of students	Degrees
Salt and Vinegar ■	8	8 × 10 = 80
Cheese and Onion ■	10	10 × 10 = 100
Ready Salted ■	12	12 × 10 = 120
Prawn Cocktail ■	4	4 × 10 = 40
Other ■	2	2 × 10 = 20

Key Point

Always align your protractor's zero line with your starting line.

Count up from zero to measure your angle.

Don't forget to label your chart.

Pictograms

- Data is represented by a picture or symbol in a **pictogram**.

Example

The pictogram shows how many pizzas were delivered by Ben in one week. Key 🍕 = 8 pizzas

Day	Mon	Tue	Wed	Thu	Fri	Sat	Sun
Pizza deliveries	🍕🍕🍕	🍕	🍕🍕	🍕	🍕🍕🍕🍕	🍕🍕	

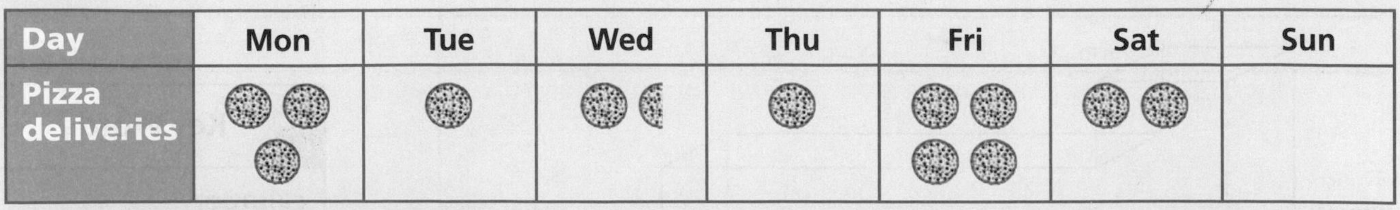

How many pizzas did Ben deliver on Friday? 8 × 4 = 32

On Sunday Ben delivered 20 pizzas. Complete the pictogram.

20 ÷ 8 = 2.5 🍕 🍕 🍕

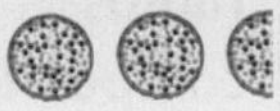

Frequency Diagrams

- **Frequency** diagrams are used to show grouped data.

Example

The heights of 100 students are shown in this table.

Height in cm	Frequency
70–80	3
81–90	4
91–100	12
101–110	24
111–120	30
121–130	22
131–140	3
141–150	2

Using the span of each height category, plot each group as a block using the frequency **axis**.

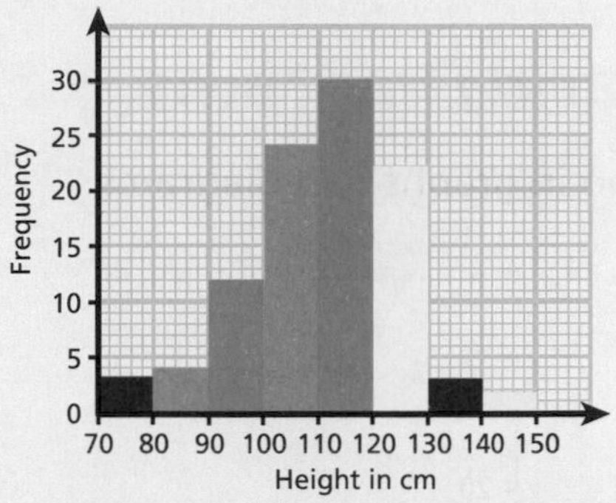

Data Comparison

- You can make comparisons using frequency diagrams.

Example

These graphs show how much time four students spend on their mobile phones in a week. Compare the two graphs.

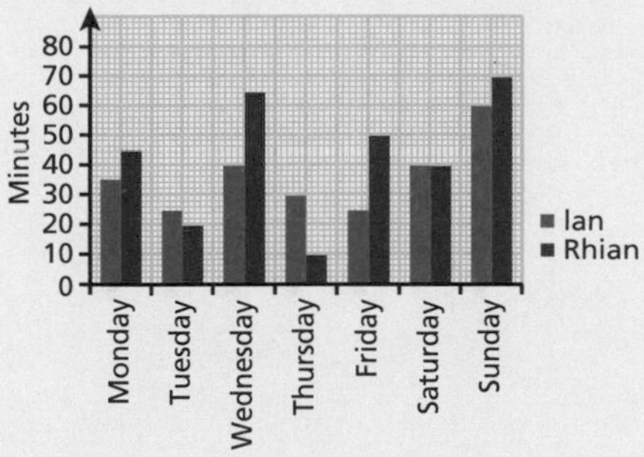

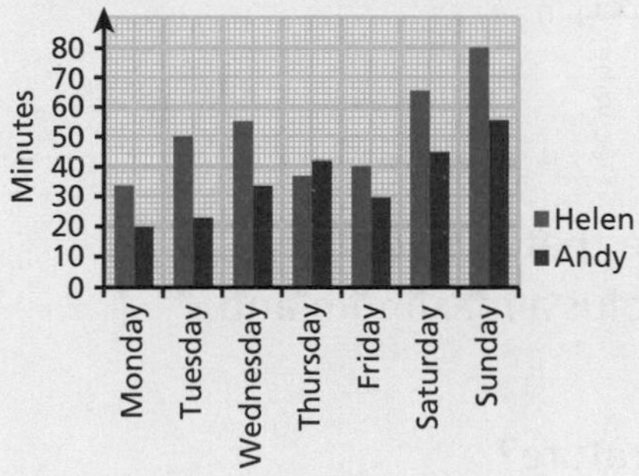

Apart from Thursday, Helen uses her phone more than Andy. In Ian's and Rhian's data, Rhian uses her phone more overall. Comparing both sets of data we can say the girls use their phones more than the boys.

Quick Test

1. If 18 people were asked a question and you were to create a pie chart to represent your data, what angle would one person be worth?
2. Using the graph above right, on which day did both Helen and Andy use their mobiles the most?
3. The following week Andy had a mean use of 40 minutes per day. Does this mean he uses his phone more than Helen now?

Key Point

Think about what is similar about the data and what is different. Look for any patterns.

Key Words

pie chart
percentage
angle
interpret
data
pictogram
frequency
axis

Statistics

Interpreting Data 2

You must be able to:

- Interpret different graphs and diagrams
- Draw a scatter graph and understand correlation
- Understand the use of statistical investigations.

Interpreting Graphs and Diagrams

- You can interpret the information in graphs and diagrams.

Example

What does this graph show?

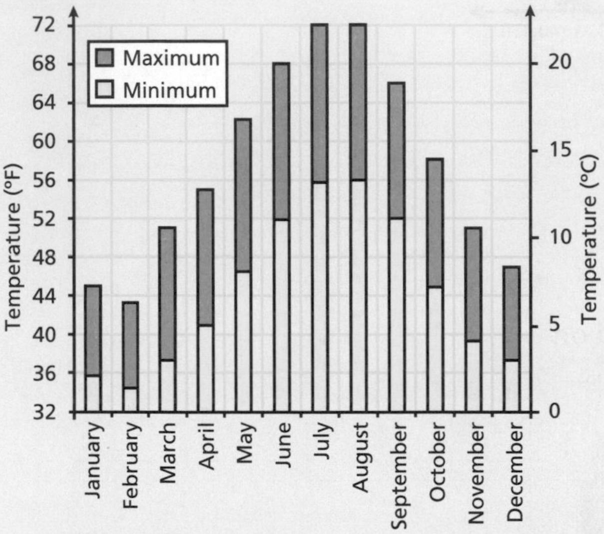

Using the labels and key, you can see that it gives the temperature and months of a year, plus a maximum and minimum temperature.

Which month has the lowest temperature?

Using the minimum temperature (yellow bars), pick the smallest bar, in this case, February.

Example

Look at this pie chart. What is the most likely way the team scores a goal?

Scoring a goal in free play represents the largest sector of the pie chart.

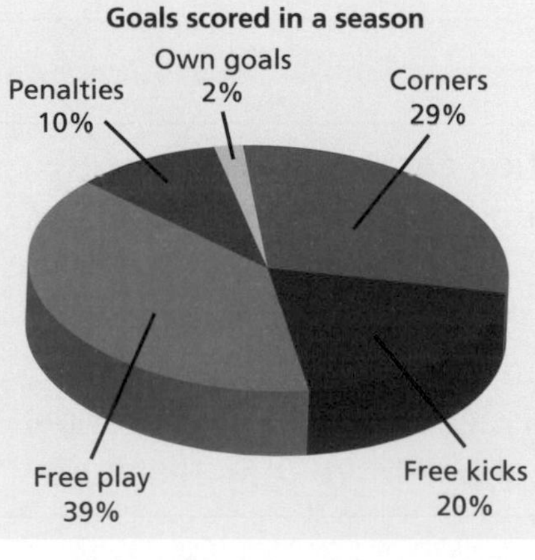

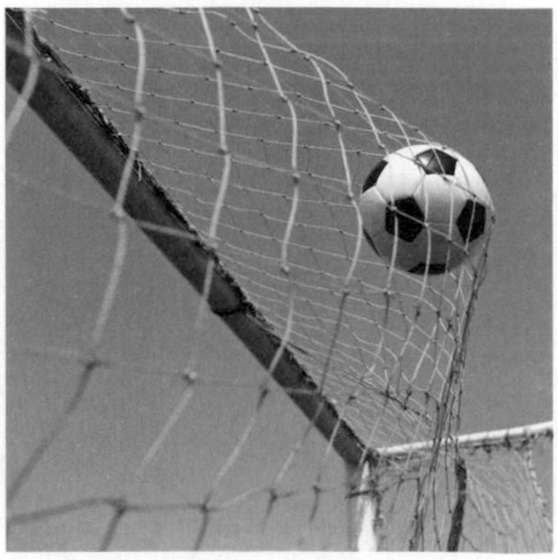

Drawing a Scatter Graph

- A **scatter graph** plots two bits of information on one graph.
- A scatter graph can show a potential relationship between data.
- You can draw a **line of best fit** through the points on a scatter graph.

Example

Here is a plot of the sales of ice cream against the amount of sun per day for 12 days. The scatter graph shows that when the weather is sunnier (and hotter), more ice creams are bought.

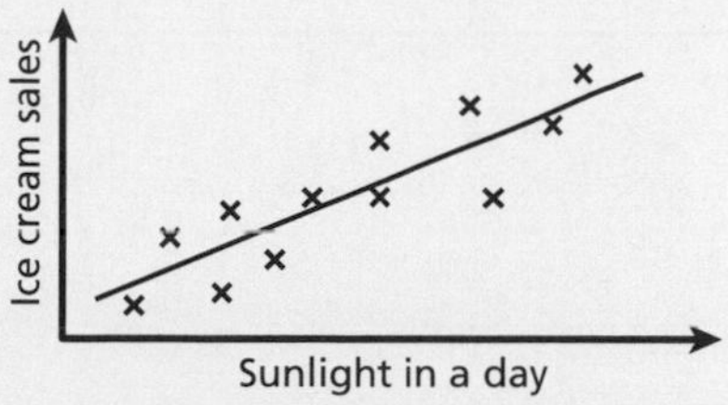

- You can describe the **correlation** and use the line of best fit to estimate data values.
- The graph above shows a positive correlation between ice cream sales and sunlight in a day.

Example

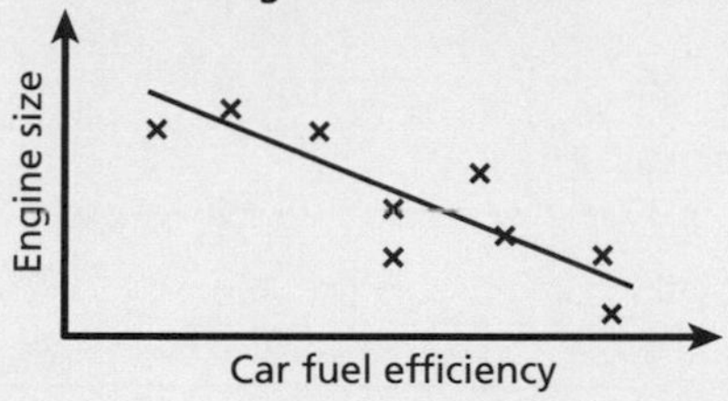

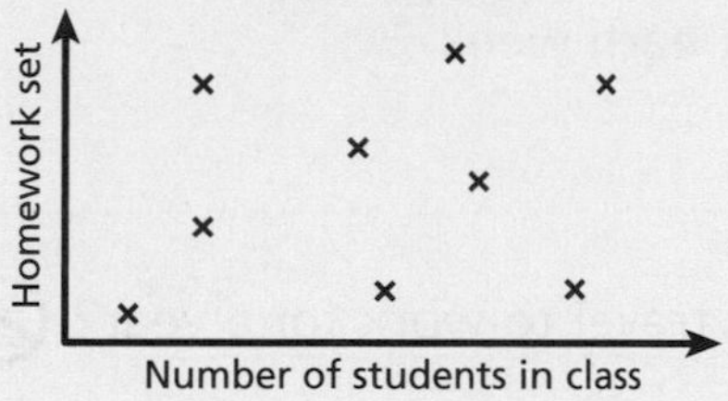

Key Point

Plot each data pair as a coordinate. A line of best fit doesn't have to start at zero.

Statistical Investigations

- Statistical investigations use **surveys** and experiments to test statements and theories to see whether they might be true or false. These statements are called **hypotheses**.
- Survey questions should be specific, timely and have no overlap on answer choices.

Key Point

Surveys should be made on a large random sample, never just a limited few.

Quick Test

1. Name two things you might plot against one another to show a positive correlation.
2. Look at the pie chart on page 48. What is the least likely way that the team scored a goal?
3. Give two ways of improving this survey question.
   How many sweets do you eat?
   None ☐ 1–2 ☐ 2–3 ☐ 3–4 ☐
4. Rewrite the question in number 3 to improve it.

Key Words

scatter graph
line of best fit
correlation
survey
hypothesis

Review Questions

Decimals

1 Work out 2.53×0.3 [2]

2 Work out $0.85 \div 0.05$ [2]

3 Circle the two numbers which add to make 9.

8.1 **0.7** **5.2** **3.8** **0.8** [1]

4 By rounding both numbers to 1 significant figure, find an estimate for the following:

$\frac{6782}{53}$ [2]

5 I pay £17.80 to travel to work each week.

I work for 48 weeks a year.

How much does it cost me to travel to work for a year? [2]

Total Marks .... / 9

1 Put the following numbers in order from smallest to biggest.

7.765, 7.675, 6.765, 7.756, 6.776 [2]

2 Thomas buys three books that cost £2.98, £3.47 and £9.54

a) How much did the books cost in total? [2]

b) How much change did he get from a £20 note? [1]

3 Use your calculator to work out $\sqrt{48}$ to 2 decimal places. [1]

4 A number has been rounded to 35 to the nearest whole number. Express the rounding error as an inequality. [1]

Total Marks .... / 7

Algebra

PS **1** Simplify these expressions:

$5k + 5 + 6k$

$k + 2 + 3k - 1$ [2]

2 Fill in the missing term in the statement below.

$3k + 4$ _______ $= k + 4$ [1]

3 Complete the missing information in the table below. The first row has been done for you.

$c \times d$	cd
$c \times d \times d$	
$c \times c \times d$	
$c \times c \times d \times d$	

[2]

Total Marks .... / 5

1 This rectangle has dimensions $a \times b$.

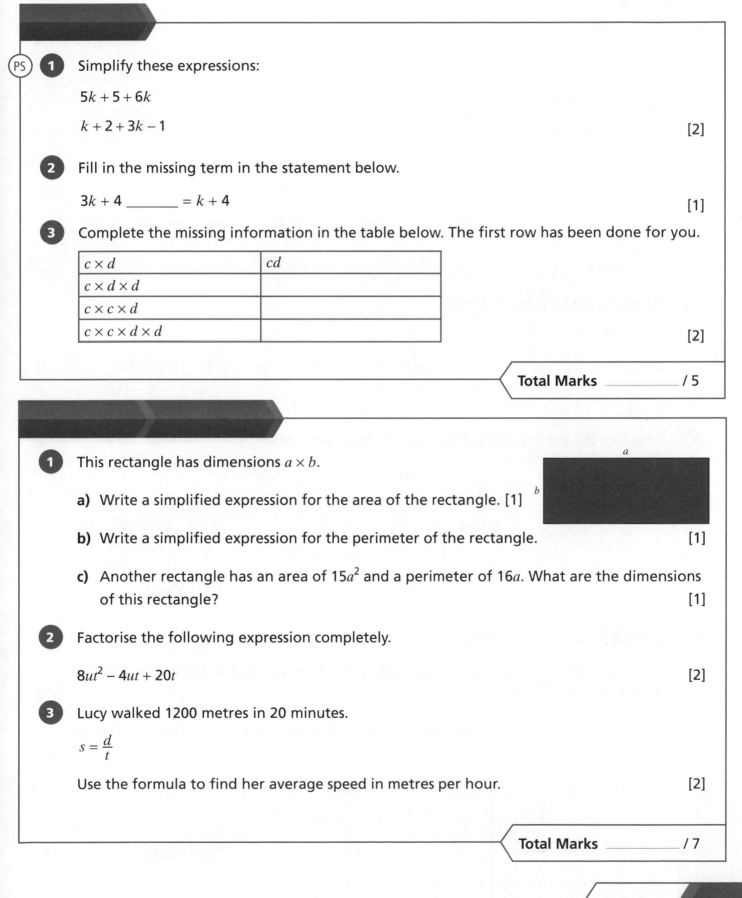

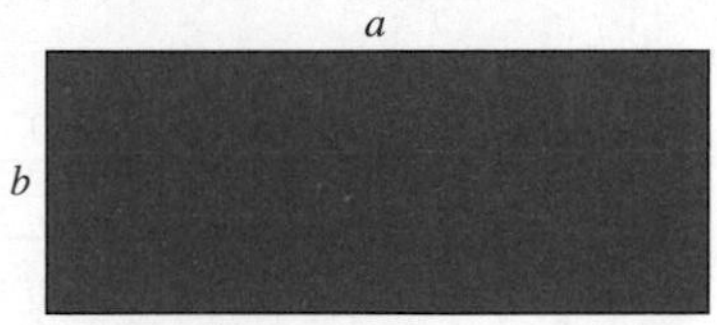

a) Write a simplified expression for the area of the rectangle. [1]

b) Write a simplified expression for the perimeter of the rectangle. [1]

c) Another rectangle has an area of $15a^2$ and a perimeter of $16a$. What are the dimensions of this rectangle? [1]

2 Factorise the following expression completely.

$8ut^2 - 4ut + 20t$ [2]

3 Lucy walked 1200 metres in 20 minutes.

$s = \frac{d}{t}$

Use the formula to find her average speed in metres per hour. [2]

Total Marks .... / 7

Practice Questions

3D Shapes: Volume and Surface Area

1 Complete the table below.

Shape	Name	No. of faces	No. of edges	No. of vertices
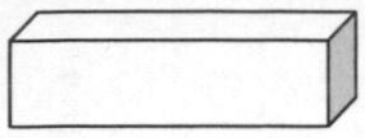				
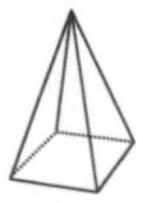				
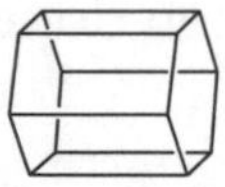				

[3]

2 Name the 3D solid for each net.

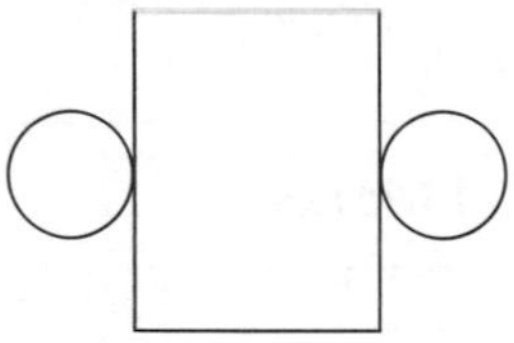

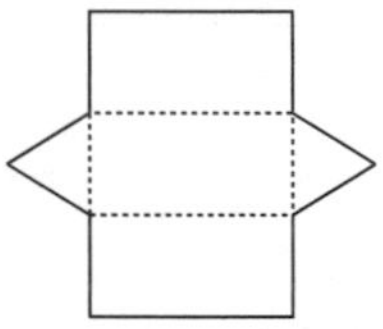

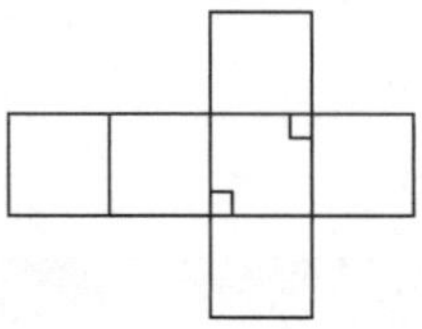

[3]

3 Calculate the volume and surface area of these two cuboids.

a)

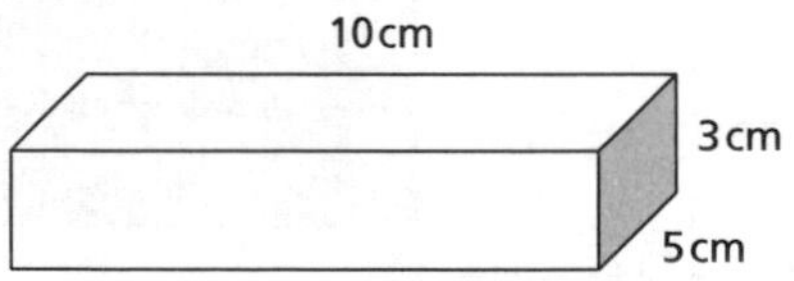

b)

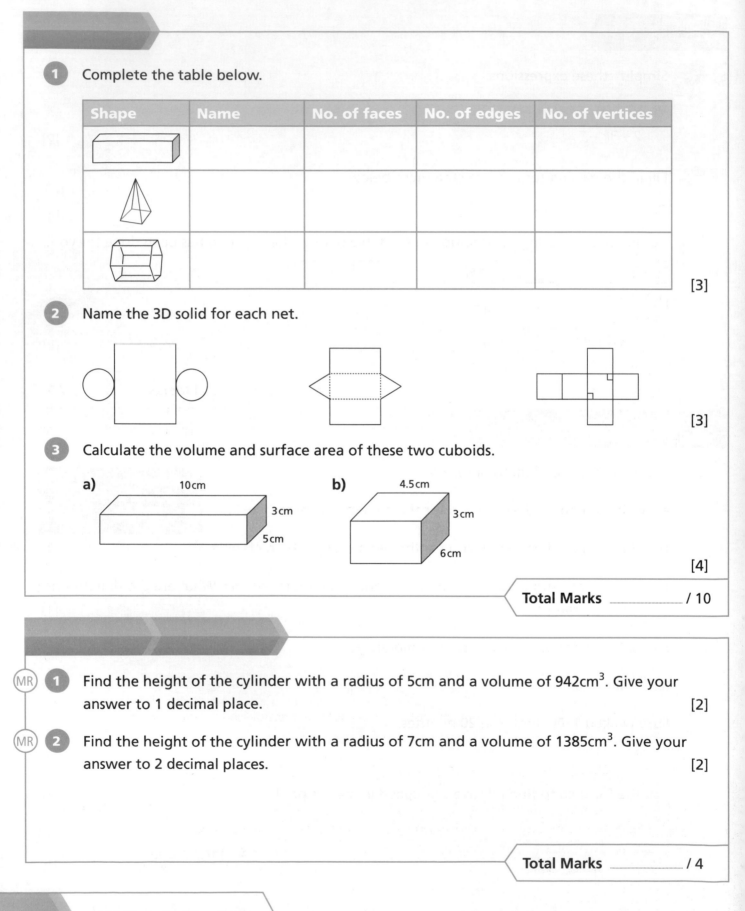

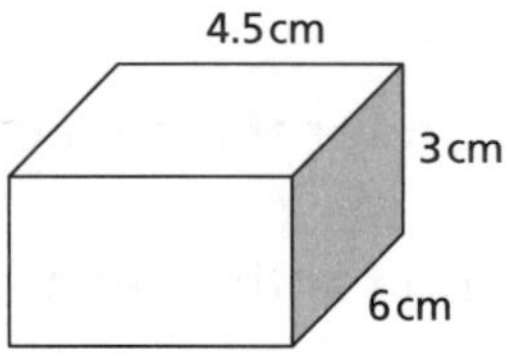

[4]

Total Marks .... / 10

MR **1** Find the height of the cylinder with a radius of 5cm and a volume of 942cm^3. Give your answer to 1 decimal place. [2]

MR **2** Find the height of the cylinder with a radius of 7cm and a volume of 1385cm^3. Give your answer to 2 decimal places. [2]

Total Marks .... / 4

Interpreting Data

MR **1** 45 people were asked what their favourite cheese was.

Complete the table and plot the information on a pie chart.

Category	Frequency	Angle
Brie	21	
Cheddar	5	
Stilton	14	
Other	5	
Total		

[6]

MR **2** The graph shows Helen and Andy's phone use during a week. On which day was their total amount of phone usage the lowest?

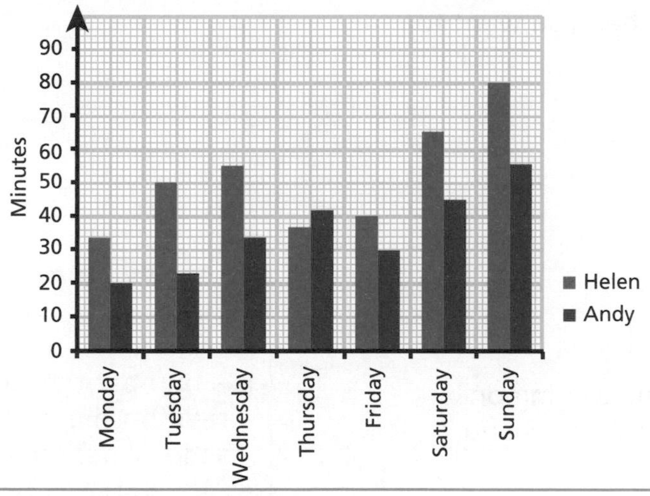

[1]

Total Marks .... / 7

1 What two things could you plot against one another to show a negative correlation? [2]

2 Write a question with response box options to determine whether people shop more over the Christmas period than at other times of the year. [4]

MR **3** **a)** Name four different types of statistical graphs or charts. [2]

b) Which one would you use to plot the information collected from asking 40 students, 'How long do you spend doing homework in a week?' Give a reason. [2]

Total Marks .... / 10

Fractions 1

You must be able to:

- Find equivalent fractions
- Add and subtract fractions.

Equivalent Fractions

- **Equivalent** fractions are fractions that are equal despite the **denominators** being different.

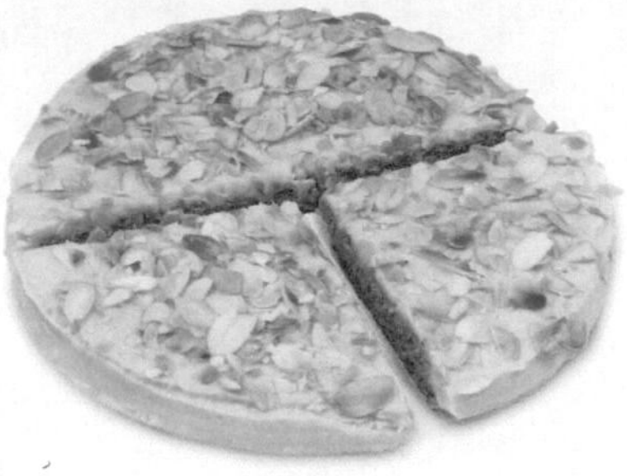

Example

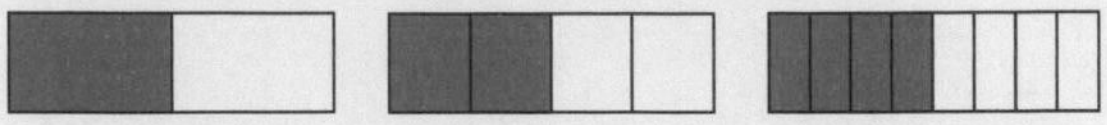

$\frac{1}{2}$ $\quad$ $\frac{2}{4} = \frac{1}{2}$ $\quad$ $\frac{4}{8} = \frac{1}{2}$

- You can create an equivalent fraction by keeping the ratio between the **numerator** and denominator the same.
- You do this by multiplying or dividing both the numerator and denominator by the same number.
- Creating equivalent fractions is very useful when you want to compare or evaluate different fractions.

Example

Which is the larger fraction, $\frac{2}{5}$ or $\frac{3}{7}$?

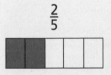

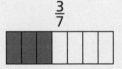

$\frac{2}{5}$ $\quad$ $\frac{3}{7}$

To compare these fractions, you need to find a common denominator – a number that appears in both the 5 and 7 times tables.

$5 \times 7 = 35$

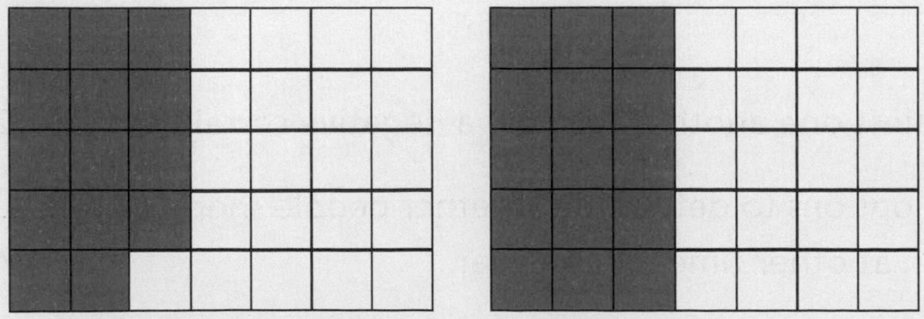

So $\frac{2}{5}$ becomes $\frac{14}{35}$ and $\frac{3}{7}$ becomes $\frac{15}{35}$

Now the denominators are equal, you can evaluate the two fractions more easily and you can see that $\frac{15}{35} = \frac{3}{7}$ is larger.

Key Point

A common denominator is a number that shares a relationship with both fractions' denominators.

For example, for 5 and 3 this would be 15, 30, 45, 60, …

Adding and Subtracting Fractions

- Adding fractions with the same denominator is straightforward. The 'tops' (numerators) are collected together.

Example

$$\frac{1}{3} + \frac{1}{3} = \frac{2}{3}$$

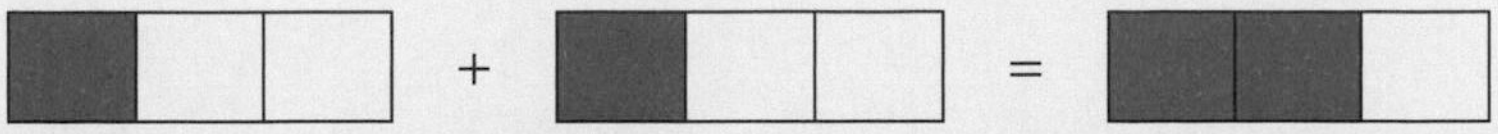

Notice the size of the 'piece', the denominator, remains the same in both the question and the answer.

- When subtracting fractions with the same denominator, simply subtract one numerator from the other.

Example

$$\frac{4}{5} - \frac{1}{5} = \frac{3}{5}$$

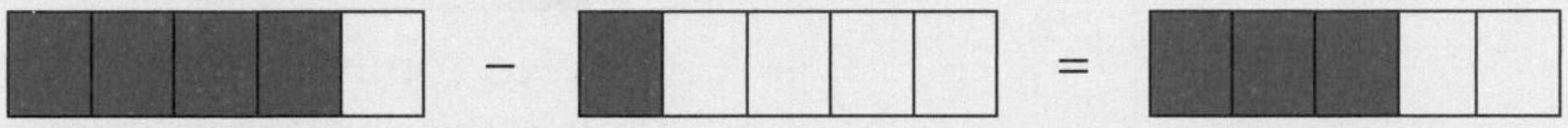

- When you have fractions with different denominators, first find equivalent fractions with a common denominator.

Example

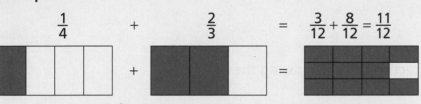

$$\frac{1}{4} + \frac{2}{3} = \frac{3}{12} + \frac{8}{12} = \frac{11}{12}$$

Here the common denominator is 12, as it is the smallest number that appears in both the 3 and 4 times tables.

This means that, for the first fraction, you have to multiply both the numerator and denominator by 3 and for the second fraction multiply them by 4.

Now the fractions are of the same size 'pieces', you can add the numerators as before.

Key Point

The size of the 'piece' (the denominator) has to be the same to perform either addition or subtraction.

Key Point

It is essential to find equivalent fractions so both fractions have the same denominator.

Quick Test

1. Find three equivalent fractions for $\frac{2}{3}$ $\frac{4}{6}$ $\frac{8}{12}$ $\frac{16}{24}$
2. Work out $\frac{2}{7} + \frac{6}{11}$ $\frac{64}{77}$
3. Work out $\frac{7}{9} - \frac{3}{8}$ $\frac{29}{72}$
4. Work out $\frac{7}{13} - \frac{1}{4}$ $\frac{15}{52}$
5. Work out $\frac{14}{25} + \frac{3}{5} - \frac{7}{20} = \frac{89}{100}$

$\frac{31}{25}$

$\frac{124}{100} - \frac{35}{100}$

Key Words

equivalent
denominator
numerator

Number

Fractions 2

You must be able to:

- Multiply and divide fractions
- Understand mixed numbers and improper fractions
- Calculate sums involving mixed numbers.

Multiplying and Dividing Fractions

- Multiplying fractions by whole numbers is not very different from multiplying whole numbers.
- The numerator is multiplied by the whole number.

Example

$\frac{2}{7} \times 3 = \frac{2}{7} + \frac{2}{7} + \frac{2}{7} = \frac{6}{7}$

- When multiplying, you multiply the 'tops' (numerators) and then multiply the 'bottoms' (denominators).

Example

$\frac{3}{5} \times \frac{2}{7} = \frac{3 \times 2}{5 \times 7} = \frac{6}{35}$

When you multiply these two fractions, it is like saying there are $\frac{3}{5}$ lots of $\frac{2}{7}$.

- When dividing, you use **inverse** operations. You change the operation to a multiplication and invert the second fraction.

Example

$\frac{2}{5} \div \frac{1}{2} = \frac{2 \times 2}{5 \times 1} = \frac{4}{5}$

When you divide fractions, you are saying how many of one of the fractions is in the other. $\frac{2}{5}$ divided into halves gives twice as many pieces.

Key Point

No common denominator is needed for multiplying or dividing fractions.

Mixed Numbers and Improper Fractions

- A **mixed number** is where there is both a whole number part and a fraction, for example $1\frac{1}{3}$
- An **improper fraction** is where the numerator is bigger than the denominator, for example $\frac{4}{3}$

Example

Change the improper fraction $\frac{14}{3}$ to a mixed number.

$$\frac{14}{3} = 4\tfrac{2}{3}$$

How many 3s are there in 14?

The remainder is left as a fraction.

Change the mixed number $4\tfrac{2}{3}$ to an improper fraction.

$$4\tfrac{2}{3} = \frac{(4 \times 3) + 2}{3} = \frac{14}{3}$$

Multiply the whole number by the denominator. 4 units is 12 thirds.

Add the $\frac{2}{3}$ to make 14 thirds.

Key Point

Improper fractions can be changed into a mixed number and vice versa.

Adding and Subtracting Mixed Numbers

- To work with mixed numbers, you first have to change them to improper fractions.
- There are four steps:
  convert to improper fractions → convert to equivalent fractions → add → convert back to a mixed number

Example

$$2\tfrac{4}{9} + 3\tfrac{1}{4} = \frac{22}{9} + \frac{13}{4} = \frac{22 \times 4}{36} + \frac{13 \times 9}{36} = \frac{205}{36} = 5\tfrac{25}{36}$$

- The same sequence of steps is needed for subtraction:
  convert to improper fractions → convert to equivalent fractions → subtract → convert back to a mixed number

Example

$$4\tfrac{4}{7} - 1\tfrac{1}{4} = \frac{32}{7} - \frac{5}{4} = \frac{32 \times 4}{28} - \frac{5 \times 7}{28} = \frac{93}{28} = 3\tfrac{9}{28}$$

Key Point

Always write the 'remainder' as a fraction.

Quick Test

Work out:

1. $\frac{4}{5} \times \frac{5}{12}$ $= \frac{20}{60} = \frac{2}{6} = \frac{1}{3}$
2. $\frac{7}{12} \div \frac{3}{7}$ $= \frac{49}{36}$
3. $\frac{12}{15} \div \frac{5}{35}$ $= \frac{420}{75}$
4. $5\tfrac{5}{6} + 3\tfrac{5}{12}$ $= \frac{111}{12}$
5. $4\tfrac{3}{10} - 2\tfrac{1}{9}$ $= \frac{577}{90}$

Key Words

inverse
mixed number
improper fraction

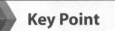

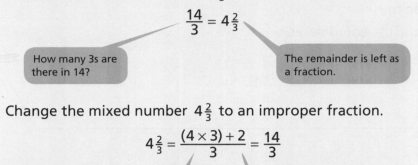

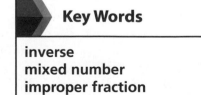

Algebra

Coordinates and Graphs 1

You must be able to:
- Understand and use coordinates
- Plot linear graphs
- Understand the components of $y = ax + b$.

Coordinates

- **Coordinates** are usually given in the form (x, y) and they are used to find certain points on a graph with an x-**axis** and a y-axis.

Example
Plot the coordinates (4, 7) and (–3, 4).

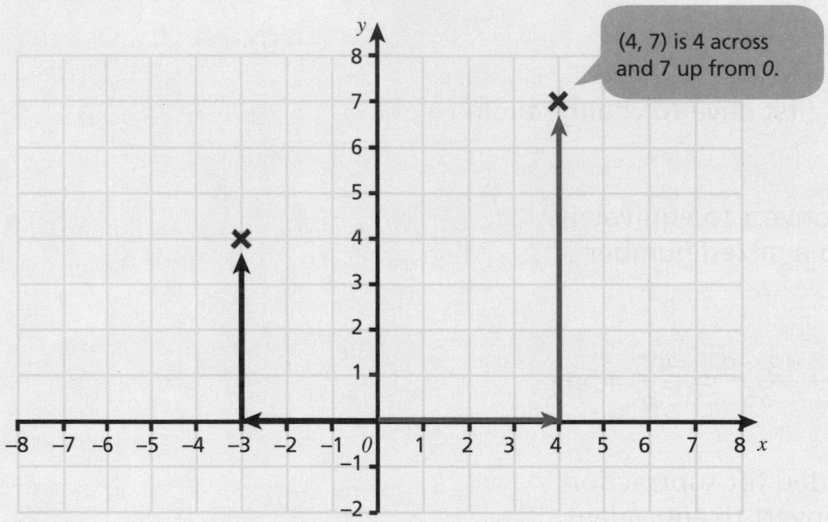

- You use the same idea to read a coordinate from a graph or chart.

Example
What are the coordinates of each corner of this square?

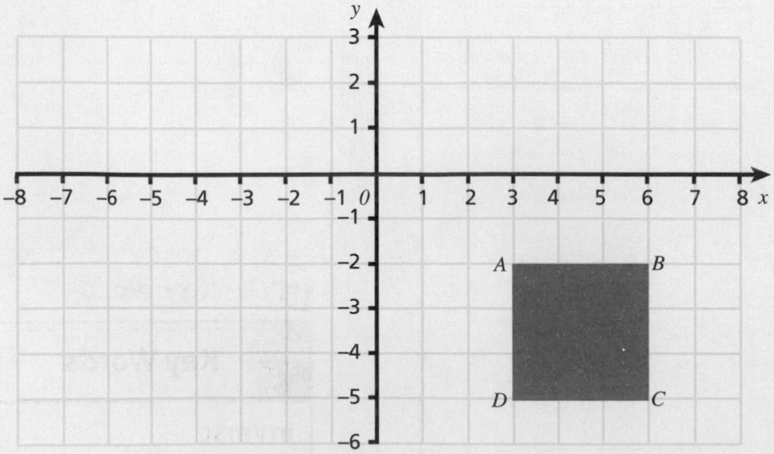

A = (3, –2), B = (6, –2), C = (6, –5), D = (3, –5)

Key Point

The x-axis is always the one that goes across the page. The y-axis always goes up the page.

When reading or plotting a coordinate, we use the first number as the position on the x-axis and the second number as the position on the y-axis.

Linear Graphs

- **Linear** graphs form a straight line.
- When plotting a graph, we need to have a rule for what we are plotting, usually an equation.

Example

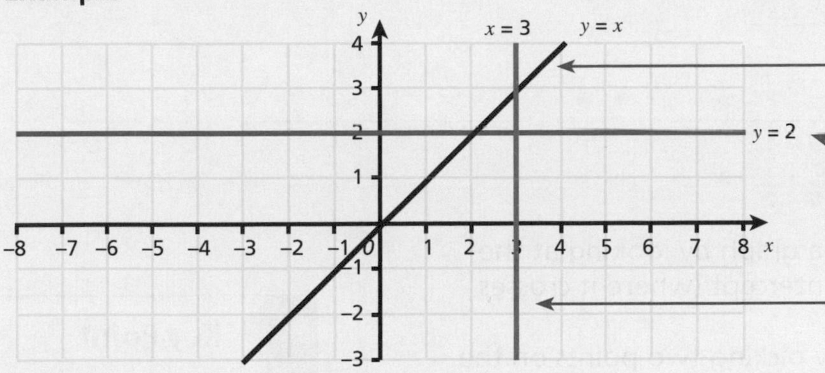

When plotting $y = x$, whatever x-coordinate you choose, the y-coordinate will be the same.

When plotting $y = 2$, you can choose any value for the x-coordinate but the y-coordinate must always equal 2.

When plotting $x = 3$, you can choose any value for the y-coordinate but the x-coordinate must always equal 3.

Graphs of $y = ax + b$

- a is used to represent the **gradient** of the graph.
- b is used to represent the **intercept**.
- To create the graph you substitute real numbers for x and y.

Example
Plot the graph $y = 2x + 1$

If $x = 1$, you can work out y: $y = 2 \times 1 + 1 = 3$
Work out other values of y by changing the value of x.

x	–1	0	1	2	3
y	–1	1	3	5	7

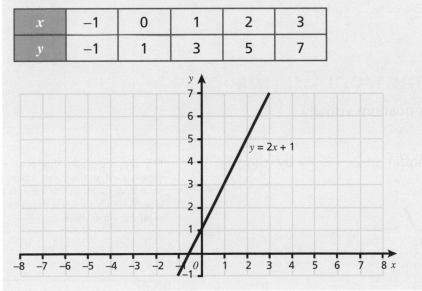

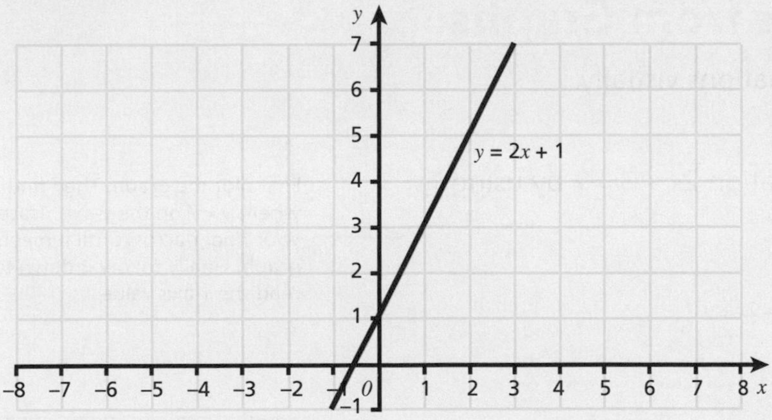

Key Point

A positive value of a will give a positive gradient. The graph will appear 'uphill'.

A negative value of a will give a negative gradient. The graph will appear 'downhill'.

Quick Test

1. Complete the table of values for $y = 3x – 5$.

x	–2	–1	0	1	2	3
y						

Key Words

coordinates
axis
linear
gradient
intercept

Coordinates and Graphs 2

You must be able to:
- Understand gradients and intercepts
- Solve equations from linear graphs
- Plot quadratic graphs.

Gradients and Intercepts

- You can work out the equation of a graph by looking at the gradient (how steep it is) and the intercept (where it crosses the y-axis).
- The gradient can be worked out by picking two points on the graph, finding the difference between the points on both the y- and x-axes and dividing them.

Key Point

$$\text{gradient} = \frac{\text{difference in } y}{\text{difference in } x}$$

Example

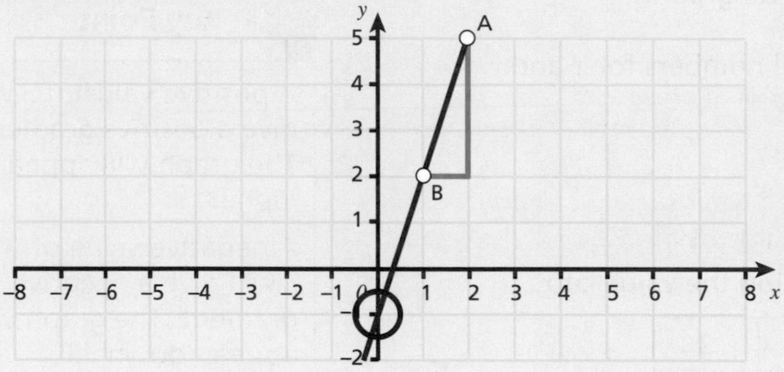

When x increases by 1, y increases by 3.

$3 \div 1 = 3$ so the gradient is 3.

The line crosses the y-axis at –1 so the intercept is –1.

The equation is $y = 3x - 1$

Solving Linear Equations from Graphs

- Graphs can be used to solve linear equations visually.

Example

You can find the solution to the equation $2x + 3 = 7$ by using the graph $y = 2x + 3$

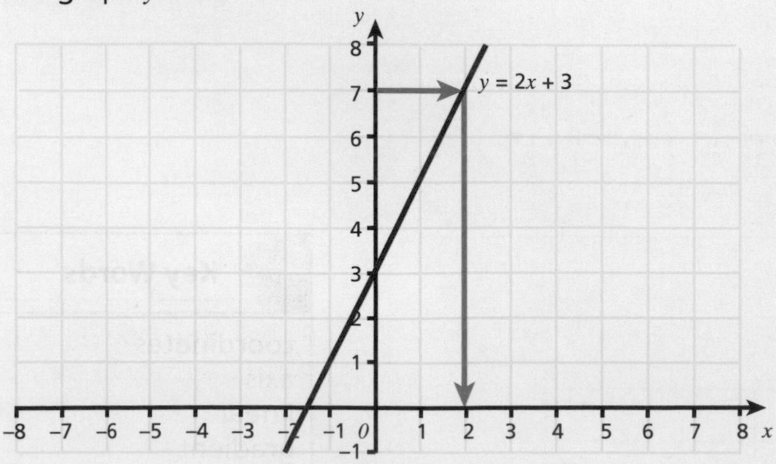

So $x = 2$

First plot the graph. Then find where $y = 7$ on the y-axis. Trace your finger across until it meets the graph. Finally follow it down to read the x-axis value.

Key Point

You can check your solution from a graph by substituting back into the equation.

For example, if $x = 2$
$y = 2x + 3 = 2(2) + 3 = 7$

- **Simultaneous equations** are two equations that are linked.
- The solution to both equations can be seen at the point where their graphs cross.

Drawing Quadratic Graphs

- **Quadratic equations** make graphs that are not linear but curved.

Example

The equation $y = x^2$ has a power in it, so this alters the graph to one that has a curve.

x	−3	−2	−1	0	1	2	3
$y = x^2$	9	4	1	0	1	4	9

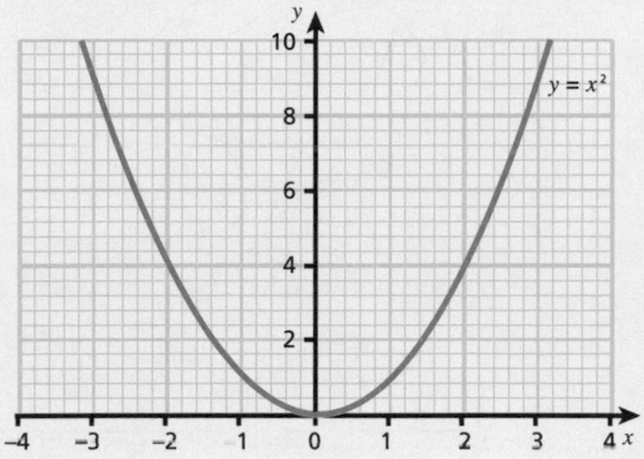

- Quadratics can take more complicated forms, but you still just **substitute** x for a real number to get the coordinates.

Example

$y = x^2 + 2x + 1$

If $x = -3$

$y = (-3)^2 + 2(-3) + 1$

$= 9 + -6 + 1 = 4$

Work out other values of y by changing the value of x.

x	−3	−2	−1	0	1	2
y	4	1	0	1	4	9

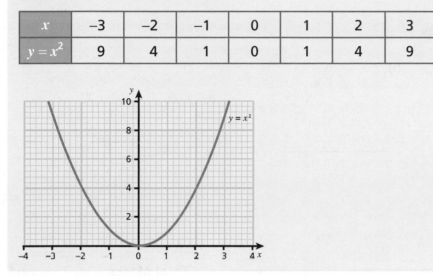

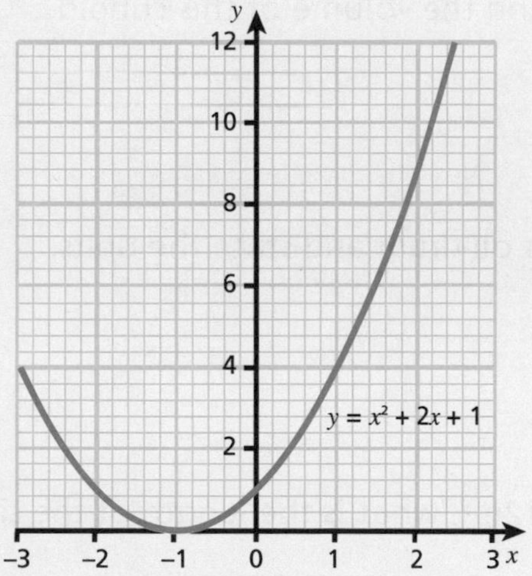

Key Point

Remember that when multiplying a negative by another negative, it becomes a positive number.

Quick Test

1. What are the gradient and intercept of these equations?
   a) $y = 3x + 5$ **b)** $y = 6x - 7$ **c)** $y = -3x + 2$
2. Fill in the table for the coordinates of $y = x^2 + 3x + 4$

x	−3	−2	−1	0	1	2	3
y							

Key Words

simultaneous equations
quadratic equation
substitute

Review Questions

3D Shapes: Volume and Surface Area

1 Name the shape that has five faces, five vertices and eight edges. [1]

2 Draw the nets of these shapes. [3]

a)

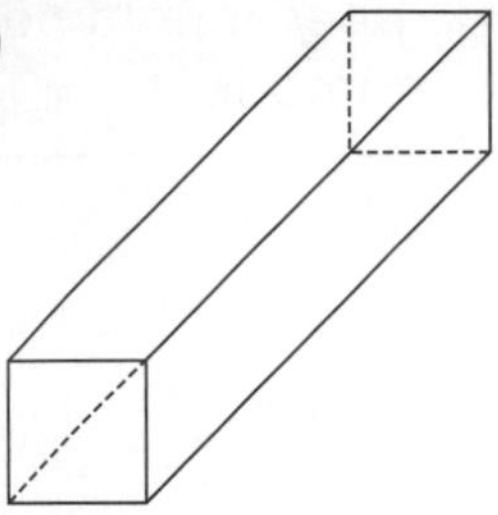

b)

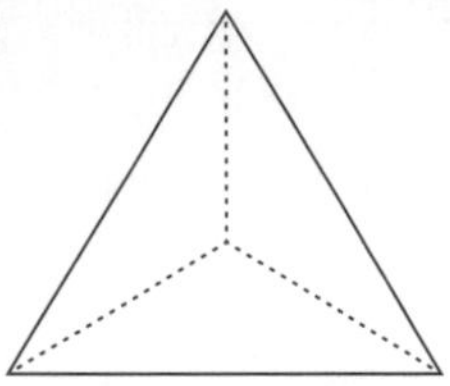

c)

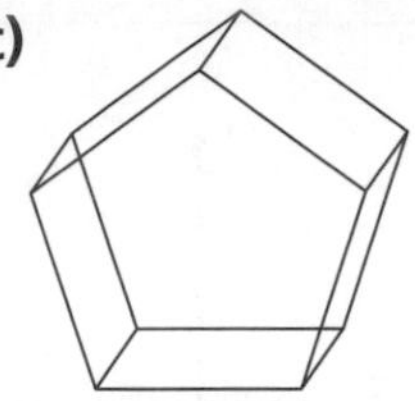

Total Marks .... / 4

1 Work out the surface area and the volume of this cuboid. Do not forget the units. [4]

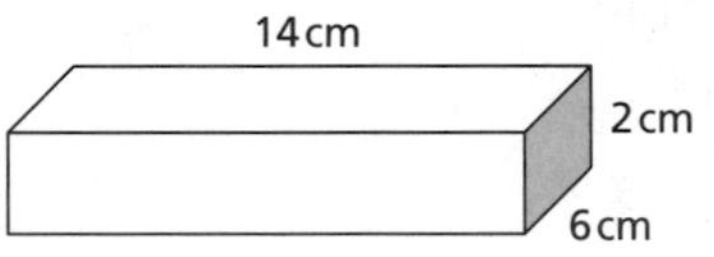

2 Work out the volume of this oil drum and state the units. [3]

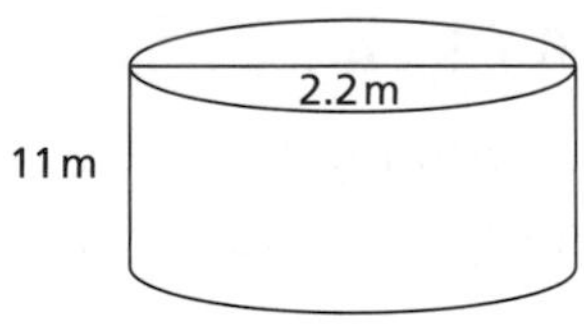

MR **3** If the volume of a cube is 512m^3, what is the length of the sides in cm? [2]

MR **4** Phil has 1100cm^2 of paper to wrap this birthday present.

Has he got enough paper? [2]

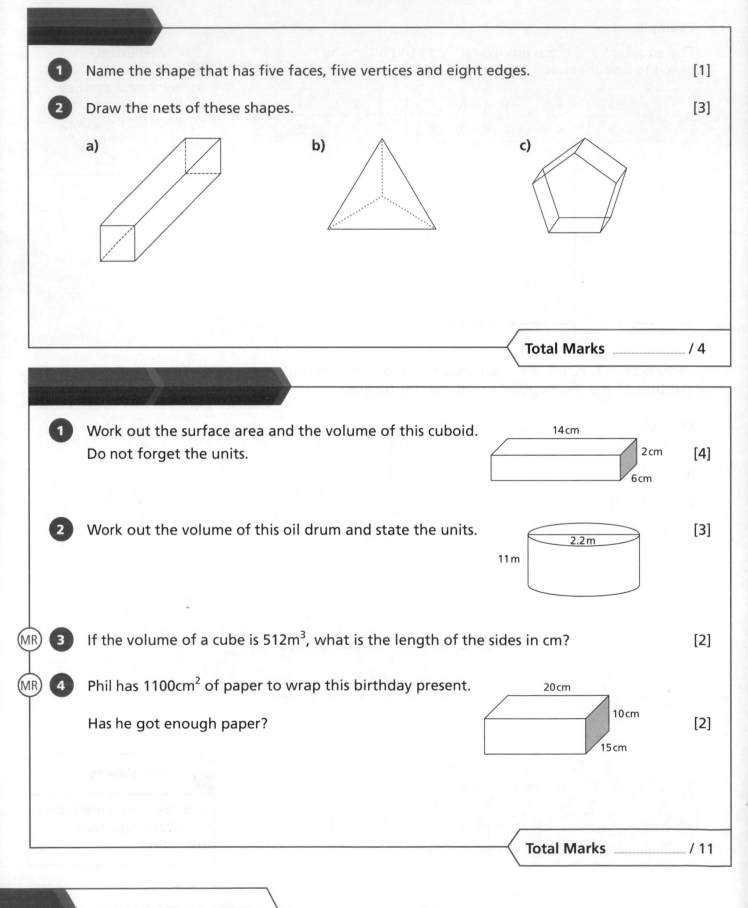

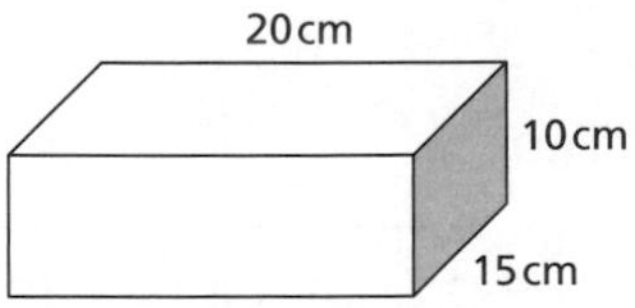

Total Marks .... / 11

Interpreting Data

1 Lalana works at a call centre. The table shows the calls she took in one day.

Call length (t minutes)	Frequency
$0 < t \leqslant 2$	25
$2 < t \leqslant 4$	40
$4 < t \leqslant 6$	18
$6 < t \leqslant 8$	10
$8 < t \leqslant 10$	4

Draw a frequency diagram for the data. [3]

Total Marks .... / 3

MR **1** Plot the following data on a scatter graph. Discuss any patterns and explain what the graph shows. [4]

TV viewing figures (in 1000s)	50	45	25	65	80	75	40	30	55
TV advert spend (in £1000s)	40	30	10	45	60	70	35	15	30

2 For each survey question below, state two things that could be improved.

a) Do you eat a lot of junk food? Yes ☐ No ☐

b) How much fruit do you eat in a week? 1 ☐ 2 ☐ 3 ☐ 4 ☐ [4]

Total Marks .... / 8

Practice Questions

Fractions

1. Prove, using the grids below, that $\frac{3}{5}$ is smaller than $\frac{2}{3}$.

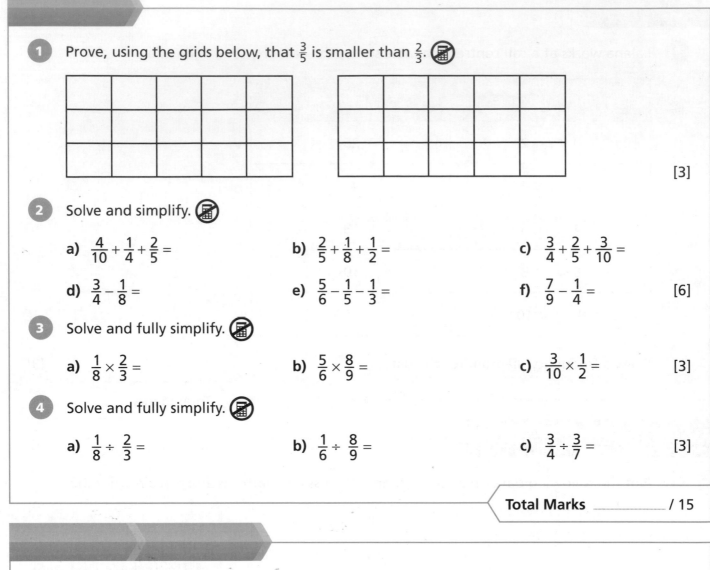

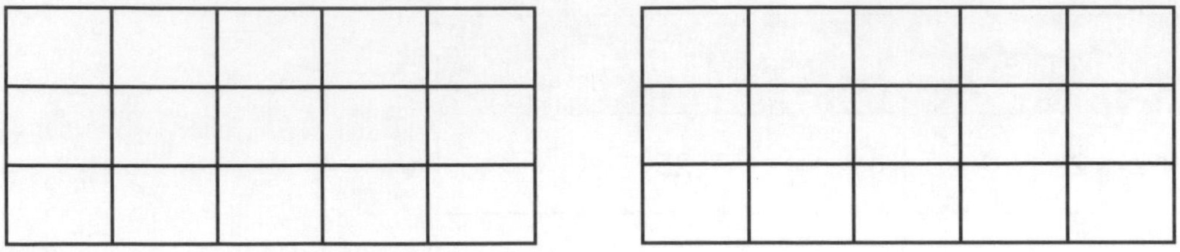

[3]

2. Solve and simplify.

a) $\frac{4}{10}+\frac{1}{4}+\frac{2}{5}=$ b) $\frac{2}{5}+\frac{1}{8}+\frac{1}{2}=$ c) $\frac{3}{4}+\frac{2}{5}+\frac{3}{10}=$

d) $\frac{3}{4}-\frac{1}{8}=$ e) $\frac{5}{6}-\frac{1}{5}-\frac{1}{3}=$ f) $\frac{7}{9}-\frac{1}{4}=$ [6]

3. Solve and fully simplify.

a) $\frac{1}{8}\times\frac{2}{3}=$ b) $\frac{5}{6}\times\frac{8}{9}=$ c) $\frac{3}{10}\times\frac{1}{2}=$ [3]

4. Solve and fully simplify.

a) $\frac{1}{8}\div\frac{2}{3}=$ b) $\frac{1}{6}\div\frac{8}{9}=$ c) $\frac{3}{4}\div\frac{3}{7}=$ [3]

Total Marks .... / 15

1. Solve each of these, giving your answer as a mixed number.

a) $4\frac{3}{8}+2\frac{1}{5}=$ [2]

b) $3\frac{3}{5}+2\frac{3}{9}=$ [2]

c) $7\frac{1}{4}-2\frac{8}{11}=$ [2]

d) $2\frac{1}{5}-1\frac{3}{7}=$ [2]

Total Marks .... / 8

Coordinates and Graphs

1. The two points on this grid are the opposite corners of a square. Find the coordinates of the missing vertices (corners).

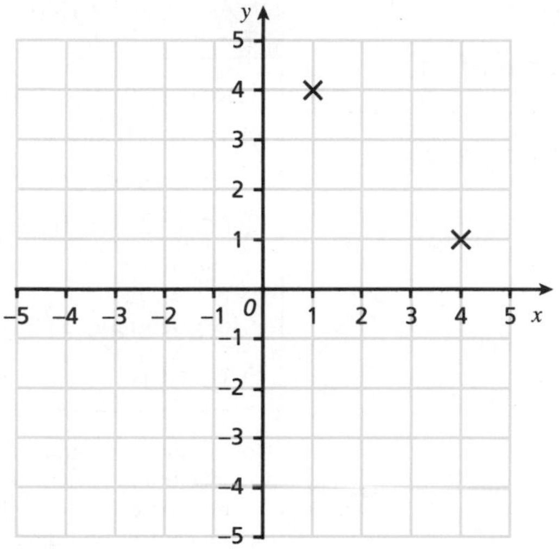

[2]

2. Draw a graph that shows the lines $x = -3$ and $y = 2$. [2]

3. Complete the table for the equation $y = 3x - 4$ and plot your results on a graph. [3]

x	–1	0	1	2	3	4
y						

Total Marks .... / 7

1. Complete the table below for the quadratic equation $y = x^2 + 3x - 2$ [3]

x	–3	–2	–1	0	1	2	3
y							

Total Marks .... / 3

Angles 1

You must be able to:

- Measure and draw angles
- Use properties in triangles to solve angle problems
- Use properties in quadrilaterals to solve angle problems
- Bisect an angle.

How to Measure and Draw an Angle

- **Angles** are measured in **degrees**, denoted by a little circle after the number, i.e. 45°.
- The piece of equipment used to measure and draw an angle is called a **protractor**.

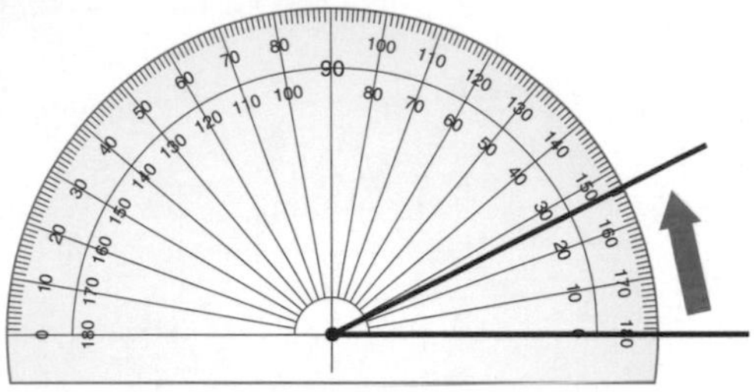

To measure the angle of a line:
- Place the protractor with the zero line on the base line.
- The centre should be level with the point where the two lines cross.
- Counting up from zero, count the degrees of the angle you are measuring.

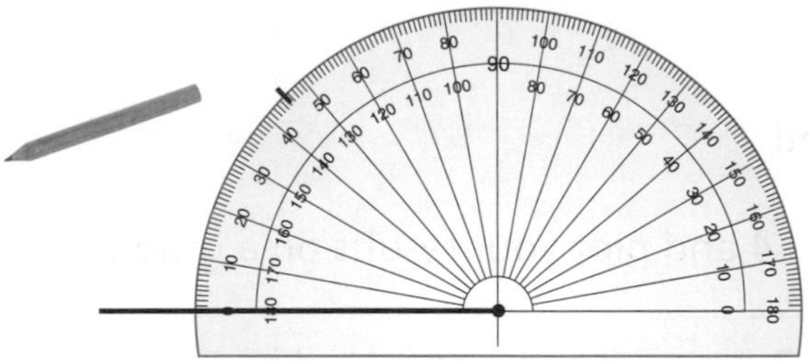

To draw an angle:
- Draw a base line for the angle.
- Line up your protractor, putting the centre on the end of the line.
- Count up from zero until you reach your angle, e.g. 45°.
- Put a mark. Remove the protractor and draw a straight line joining the end of the base line to your point.

Key Point

There are two scales on your protractor. This is so the protractor can be used in two different directions.

Always start at zero and count up.

Angles in a Triangle

- Angles in any **triangle** add up to 180°. You can use this fact to help you work out missing angles.
- An isosceles triangle has two sides the same and two (base) angles the same.
- A right-angled triangle has a 90° angle, so if you have one more angle you can work out the remaining one.

Example

Find the missing angle x in these triangles.

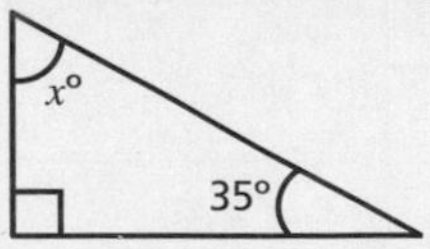

$180° - (90° + 35°) = x° = 55°$

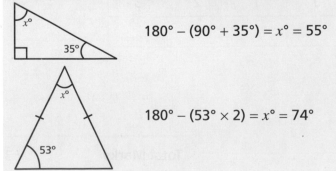

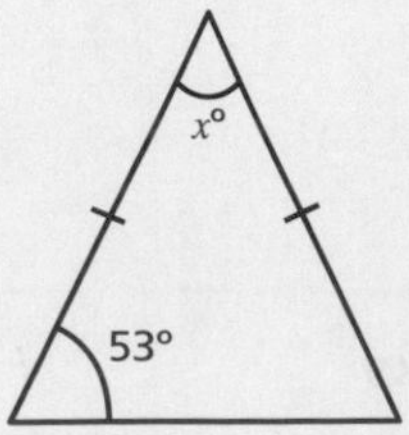

$180° - (53° \times 2) = x° = 74°$

Angles in a Quadrilateral

- Angles in any **quadrilateral** add up to 360°.

Example

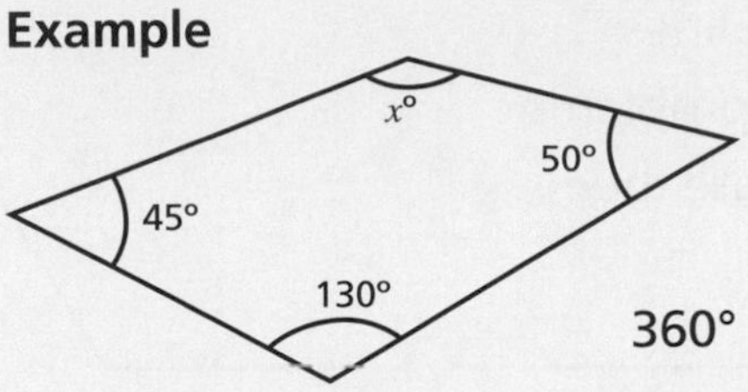

$360° - (50° + 45° + 130°) = x° = 135°$

- A parallelogram has two sets of equal angles. The opposite angles are equal.
- Only one angle is needed to be able to work out the others.

Example

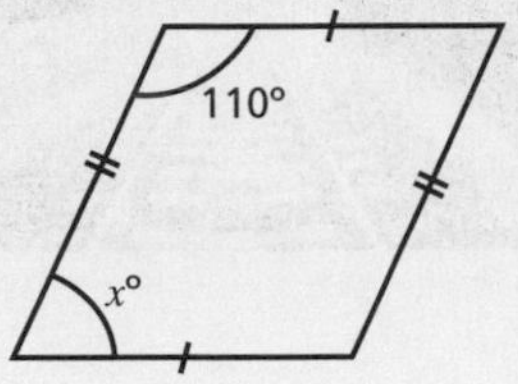

$360° - (110° \times 2) = 140°$
Now share this angle equally between the remaining two corners:
$x = 140° \div 2 = 70°$

> **Key Point**
>
> Remember that each triangle or quadrilateral will present properties that are useful when working out missing angles.

Bisecting an Angle

- **Bisect** means to cut exactly into two.

Example

1. Open your compass to a distance that is at least halfway along one of the lines.
2. Put the point of the compass where the two lines touch. Mark an arc on both lines.
3. Now place the point on one of these arcs and draw another arc between the lines.
4. Repeat for the arc on the other line, creating a cross.
5. Draw a line from where the lines touch through the cross.

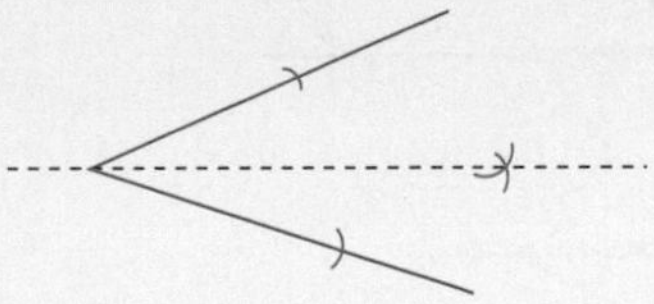

Quick Test

1. Using a protractor draw an angle of:
   a) 48° b) 84° c) 125° d) 167°
2. Find the missing angle in each of these shapes.

a) b) c)

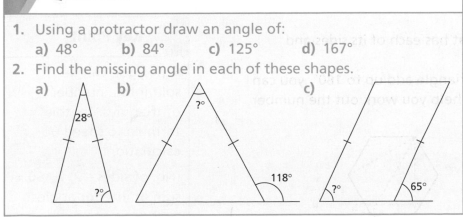

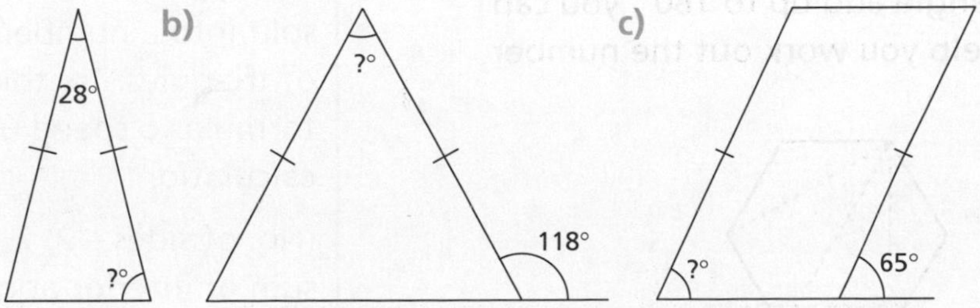

> **Key Words**
>
> **angle**
> **degree**
> **protractor**
> **triangle**
> **quadrilateral**
> **bisect**

Angles 2

You must be able to:

- Understand and calculate angles in parallel lines
- Use properties of a polygon to solve angle problems
- Use properties of some polygons to tessellate them.

Angles in Parallel Lines

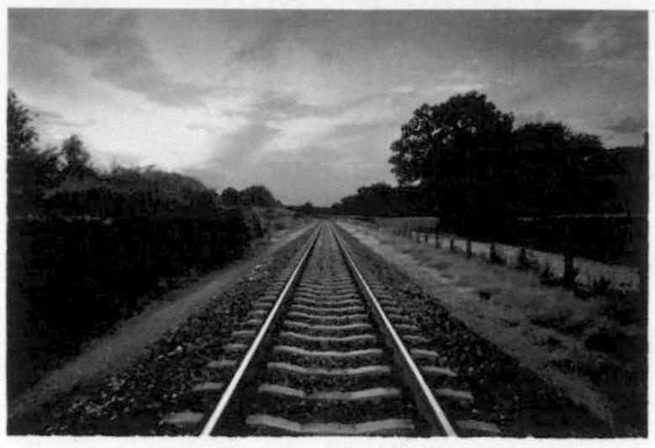

- **Parallel lines are lines that run at the same angle.**
- **Using parallel lines and a line that crosses them, you can apply some observations to help find missing angles.**
- **The angles represented by  are equal. They are called corresponding angles.**
- **The angles represented by  are also equal. They are called alternate angles.**

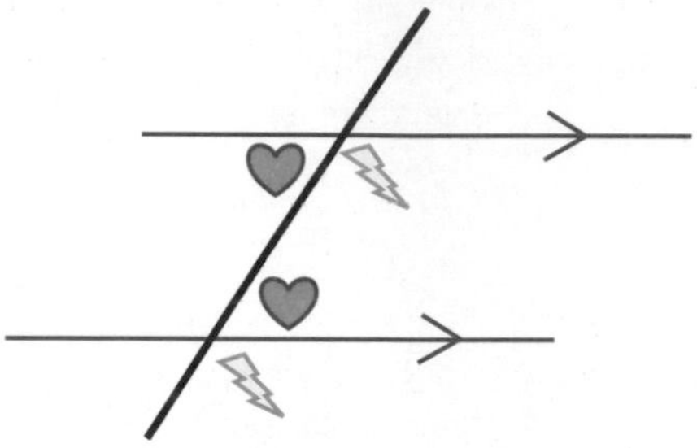

- **Alternate angles are also called Z angles, because the lines look like a Z.**
- **Corresponding angles are also called F angles, because the lines look like an F.**

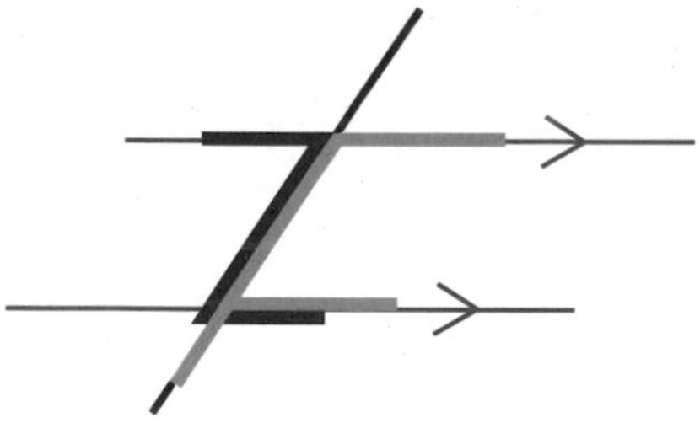

Angles in Polygons

- **A regular polygon is a shape that has each of its sides and angles equal.**
- **Using the fact that angles in a triangle add up to 180°, you can split any shape into triangles to help you work out the number of degrees inside that shape.**

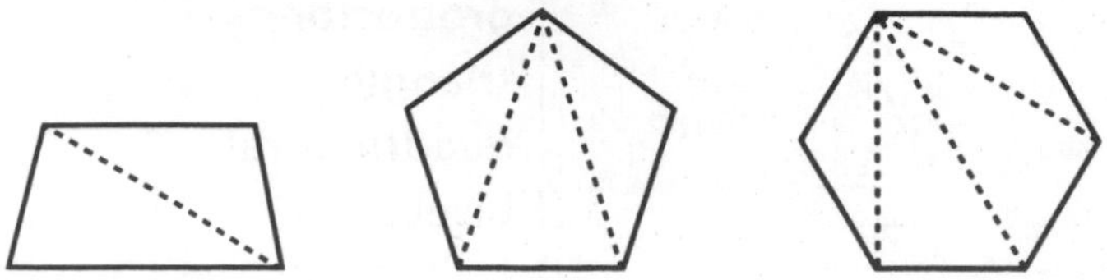

Key Point

A polygon can be split into a number of triangles. Try this formula to speed up calculation:

(No. of sides – 2) × 180 = sum of interior angles

- Dividing the total number of degrees inside a regular polygon by the number of vertices will give the size of one interior angle.
- The sum of the exterior angles of any shape always equals 360°.

Shape	Number of sides	Sum of the interior angles
Triangle	3	180°
Quadrilateral	4	360°
Pentagon	5	540°
Hexagon	6	720°
Heptagon	7	900°
Octagon	8	1080°

Example

A regular polygon has interior angles of 140°. How many sides does it have?

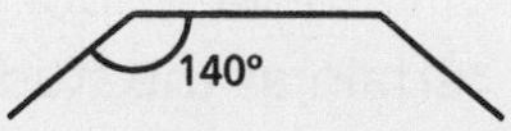

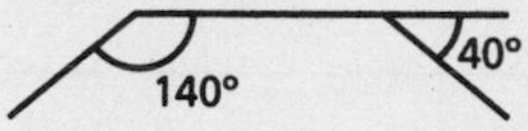

The exterior angle of each part of the polygon is 180° – 140° = 40°

360° ÷ 40° = 9, so the polygon has nine sides.

Polygons and Tessellation

- Tessellation is where you repeat a shape or a number of shapes so they fit without any overlaps or gaps.

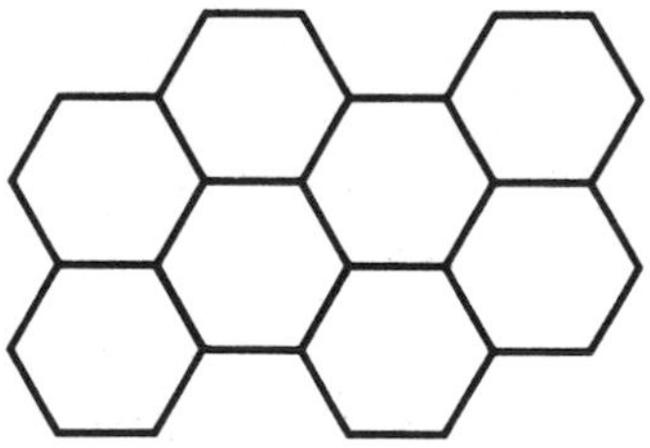

Key Point

You may notice that shapes that tessellate have angles that add up to 360° about one point.

Quick Test

1. Find the missing angles:

a) 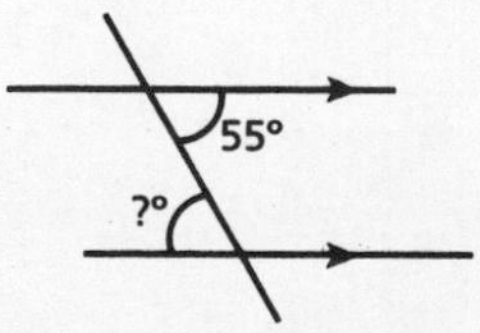

b) 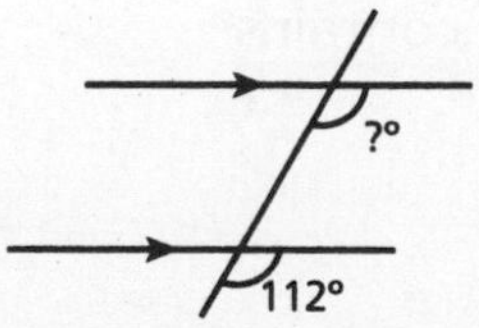

c) 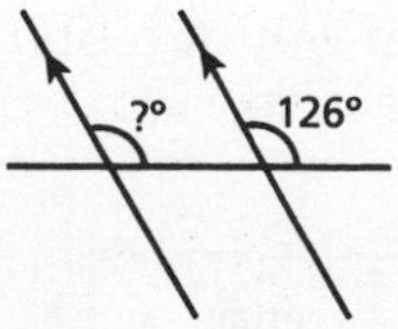

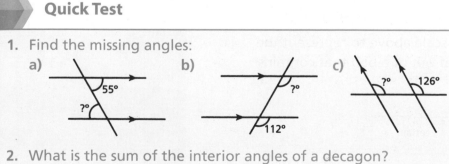

2. What is the sum of the interior angles of a decagon?
3. Name a regular shape that would tessellate.

Key Words

parallel
corresponding
alternate
regular polygon
interior angle
exterior angle
tessellation

Probability 1

You must be able to:

- Recognise and use words associated with probability
- Construct and use a probability scale
- Calculate the probability of an event not occurring
- Construct and use sample spaces.

Probability Words

- Certain words are used to describe the **probability** of an **event** happening. How would you describe:
  - The probability of there being 40 days in a month?
    Impossible – there are at most 31 days in a month.
  - The probability that a student will attend school tomorrow?
    Likely – it cannot be said to be certain as the student might be on school holiday or ill and not attending school.
  - The probability that I roll a 2 on a dice?
    Unlikely – there are six possible outcomes and the number 2 is only one of these.
  - The probability that out of a bag with only green sweets in, a green sweet is pulled out?
    Certain – in this case there is no other outcome possible.
  - The probability a coin is tossed and it lands on heads?
    An **even chance** – the event outcome could be a head or a tail, two equally likely options.

Key Point

Try to consider the event with all the possible outcomes.

Once all the possible outcomes have been considered, the word can be selected.

Probability Scale

- A **probability scale** is a way of representing the probability words visually – putting the words we used above on a number line.
- The two extremes go at either end of the scale and the other probability words can be slotted in between:

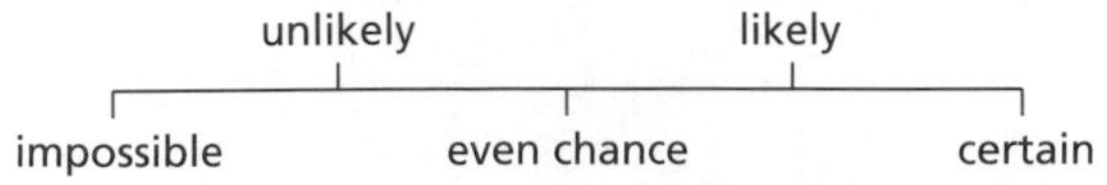

Example

Put an arrow on the probability scale above to represent the probability of pulling a red sweet out of a bag that contains 7 red sweets and 1 blue sweet.

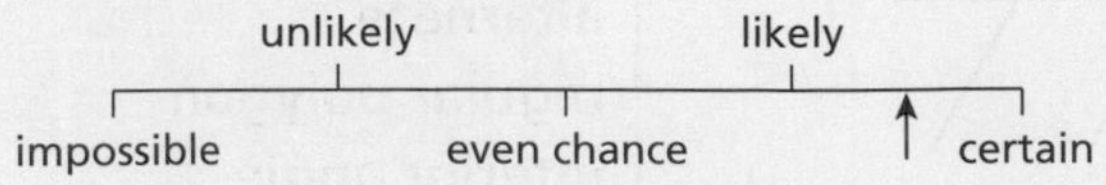

- If an event has a more than equal outcome, then it is said to be **biased**.

Probability of an Event Not Occurring

- Probability can be thought of in terms of numbers as well as words. You could think of something happening as a fraction or a decimal.
- The probability of an event not happening can be summarised as 1 – event occurring.

Example

A non-biased six-sided dice is rolled:

P(rolling a 1) = $\frac{1}{6}$

P(rolling a 2) = $\frac{1}{6}$

P(rolling a 3) = $\frac{1}{6}$

P(rolling a 4) = $\frac{1}{6}$

P(rolling a 5) = $\frac{1}{6}$

P(rolling a 6) = $\frac{1}{6}$

P(not rolling a 4 or 5) = $1 - (\frac{1}{6} + \frac{1}{6}) = \frac{4}{6} = \frac{2}{3}$

Example

The probability of the weather being cloudy = 0.4

So the probability of it *not* being cloudy is 1 – 0.4 = 0.6

Key Point

When you add up all the possible event outcomes, they should add up to 1.

Use this fact to help you work out the probability of an event **not** occurring, by adding all the possible outcomes of the event occurring and taking them away from 1.

Sample Spaces

- A **sample space** shows all the possible outcomes of the event.

Example

The sample space for flipping a coin and rolling a dice is:

H1	H2	H3	H4	H5	H6
T1	T2	T3	T4	T5	T6

This shows all the possible outcomes and helps us to calculate the probability of events occurring.

Quick Test

1. How can the probability of rain in Manchester in October be described?
2. Show the event in question 1 on a probability scale.
3. A bag contains 5 red sweets and 10 blue sweets.
   a) What is the probability of picking a red sweet?
   b) What is the probability of not picking a red sweet?
4. If it rains 0.75 of the time in the rainforest and is cloudy 0.1 of the time, what is the probability it isn't raining or cloudy?

Key Words

probability
event
impossible
likely
unlikely
certain
even chance
probability scale
biased
sample space

Probability 2

You must be able to:

- Understand what mutually exclusive events are
- Calculate a probability with and without a table
- Work with experimental probability
- Understand Venn diagrams and set notation.

Mutually Exclusive Events

- **Mutually exclusive** events can be defined as things that can't happen at the same time.
- For example, this spinner can't land on yellow and blue at the same time as it only has one pointer. You can say that the probability of getting a yellow and a blue is 0, i.e. mutually exclusive.
- You could calculate the probability of spinning a yellow or a blue, which would be $\frac{1}{4} + \frac{1}{4} = \frac{1}{2}$
- Wearing one orange sock and one purple sock would not be mutually exclusive as these events could happen at the same time.

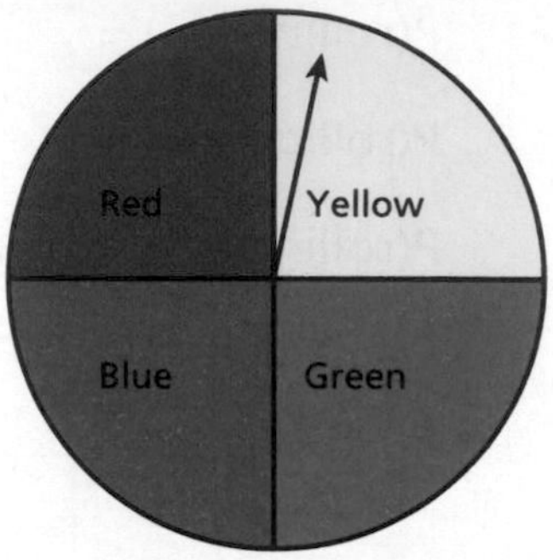

Calculating Probabilities and Tabulating Events

- Probability can be calculated by evaluating **all** outcomes.

Examples

What is the probability of picking a red ball from the bag?

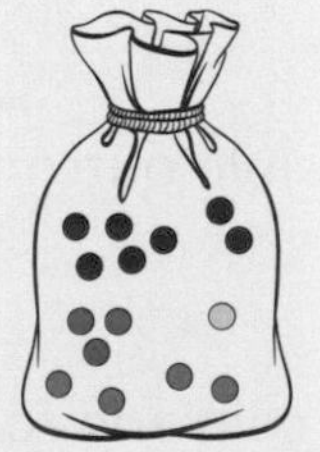

The number of red balls / The total number of balls → $\frac{7}{15}$

What is the probability of picking out a green *or* a blue ball?

$\frac{3}{15} + \frac{4}{15} = \frac{7}{15}$

There are three different colours in this bag of 15 balls: 3 yellow, some red and some blue balls. Use the table to work out the number of each colour ball.

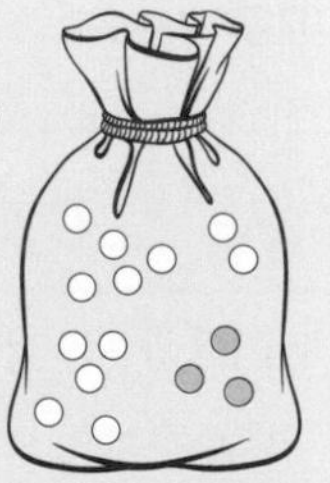

Yellow	Red	Blue
0.2	0.6	?

The probability that any ball is chosen is 1 so we can calculate the probability of drawing a blue ball: 1 – 0.6 – 0.2 = 0.2

One way of working out the actual number of each colour is to multiply the probability by the total number of balls:

15 × 0.6 = 9 (red)   0.2 × 15 = 3 (blue)

Key Point

Using the word **or** in probability usually implies that we add the probabilities of the events involved in the statement together.

Using the word **and** in probability can imply that we multiply the probabilities of the events.

Experimental Probability

- The probability of getting a 6 when rolling a six-sided dice is $\frac{1}{6}$. If you rolled the dice six times would you definitely get a 6? You may not get a 6 in six rolls, however the more times you roll the dice the more likely you are to get closer to $\frac{1}{6}$. This is **experimental probability.**

Example
Hayley sat outside her school and counted 25 cars that went past. She noted the colour of each car in this table.

Yellow	1
Red	6
Blue	4
Black	9
White	5

a) What is the probability of the next car going past being white? $\frac{5}{25} = \frac{1}{5} = 0.2$

b) How many black cars would you expect if 50 cars go past?
$\frac{9}{25} = 0.36 \qquad 0.36 \times 50 = 18$

Key Point

You are using probabilities to guess events. These don't map exact outcomes but should be used to guide you to likely and less likely events.

Venn Diagrams and Set Notation

- Venn diagrams can be used to organise sets and find probabilities.

Example
Set A is the children who like green beans. Set B is the children who like carrots.

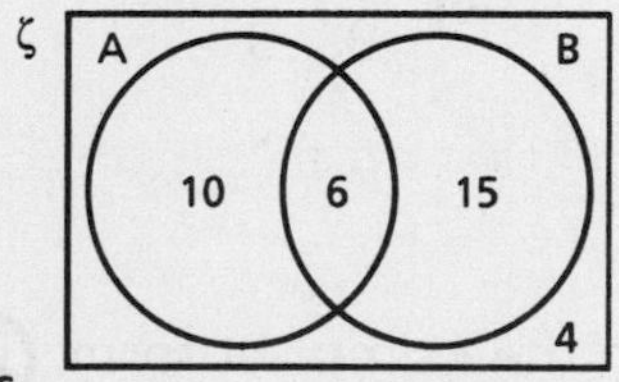

(A) = 10 + 6 = 16 children like green beans

(B) = 6 + 15 = 21 children like carrots

$(A \cup B)$ = 10 + 6 + 15 = 31 children like at least one. ← This is the union of A and B.

$(A \cap B)$ = 6 children like both green beans and carrots. ← This is the intersection of A and B.

$(A \cup B)'$ = 4 children like neither green beans nor carrots. ← The symbol ′ means not in A or B.

Quick Test

1. A bag has 20 balls of four different colours.
   a) Complete the table.

Yellow	?
Red	0.3
Green	0.2
Blue	0.1

   b) What is the probability of not getting a yellow ball?
   c) If I picked out and replaced a ball 60 times, what is the expected probability that I would pick out a green ball?

Key Words

mutually exclusive
experimental probability

Review Questions

Fractions

1 What fraction of each shape is shaded in? Give two other equivalent fractions for each. [3]

a) 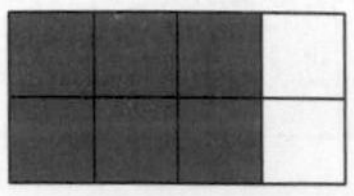

b) 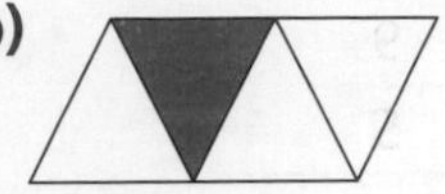

c) 

Total Marks .... / 3

1 Solve, giving your answers in their simplest form.

a) $\frac{2}{5} + \frac{1}{10} =$

b) $\frac{7}{12} + \frac{1}{4} =$

c) $\frac{1}{6} + \frac{1}{5} =$

d) $\frac{2}{7} + \frac{3}{10} =$

e) $\frac{8}{9} - \frac{1}{3} =$

f) $\frac{7}{11} - \frac{1}{2} =$

g) $\frac{9}{10} - \frac{2}{3} =$ [7]

2 Solve, giving your answers in their simplest form.

a) $\frac{4}{9} \times \frac{1}{5} =$

b) $\frac{3}{7} \times \frac{3}{10} =$

c) $\frac{5}{12} \times \frac{2}{3} =$

d) $\frac{2}{9} \div \frac{1}{4} =$

e) $\frac{4}{5} \div \frac{6}{11} =$ [5]

FS MR 3 Sally won some money in the lottery. She gave $\frac{2}{5}$ to her husband and $\frac{1}{4}$ to her daughter.

What fraction did she keep? [3]

MR 4 Keith was baking a cake. His recipe was $\frac{4}{9}$ flour, $\frac{1}{3}$ sugar and butter, and the rest was an equal split of chocolate and eggs.

What fraction does the chocolate part represent? [3]

PS 5 Change the following mixed numbers to improper fractions.

a) $8\frac{5}{9}$

b) $3\frac{2}{7}$

c) $1\frac{3}{11}$ [3]

Total Marks .... / 21

Coordinates and Graphs

1 Copy the grid shown and plot the following points:

a) (1, 2) [1]

b) (–4, 5) [1]

c) (3, –2) [1]

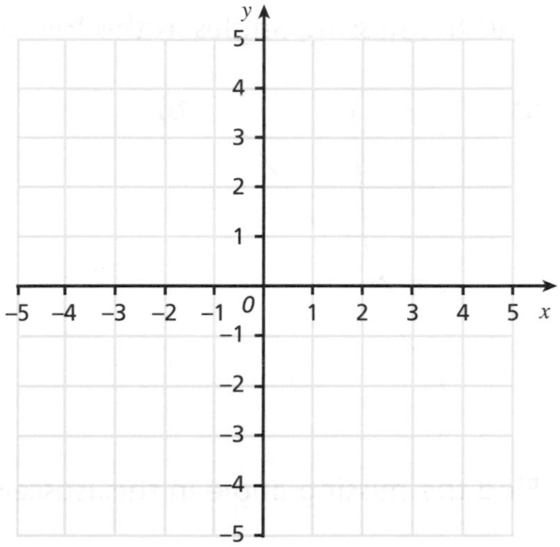

2 On the same grid plot the following lines:

a) $x = 4$ **b)** $y = -1$ [2]

Total Marks .... / 5

1 Copy and complete the arrows and the table below for the equation $y = 2x - 4$. Then plot your results on a grid like the one shown.

× → – →

x	–1	0	1	2	3
y					

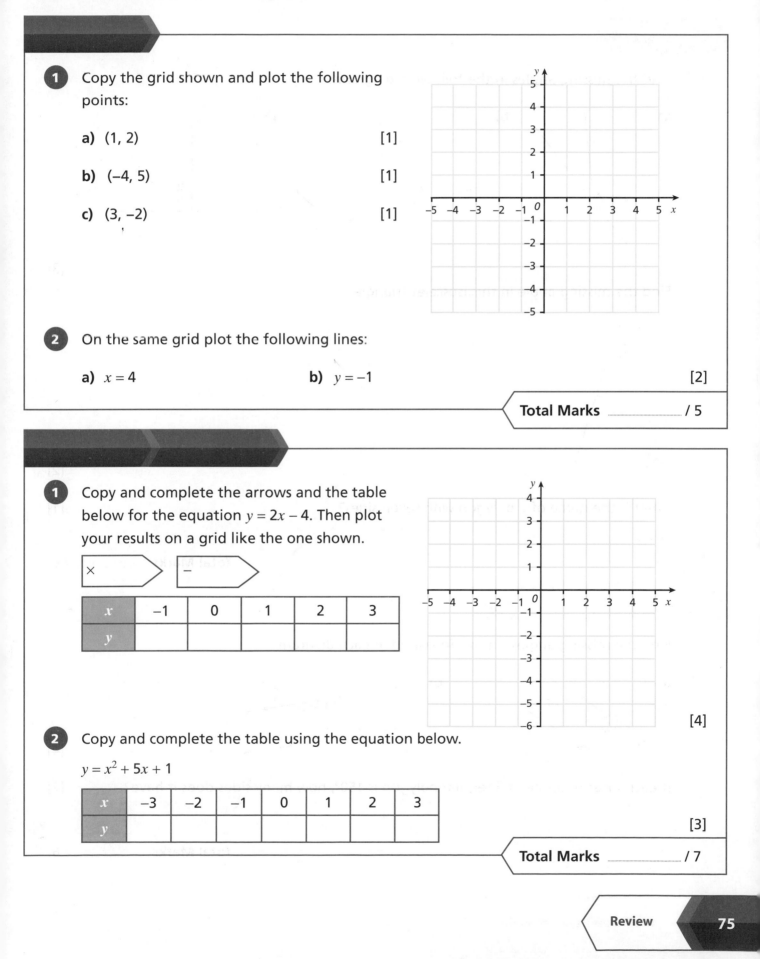

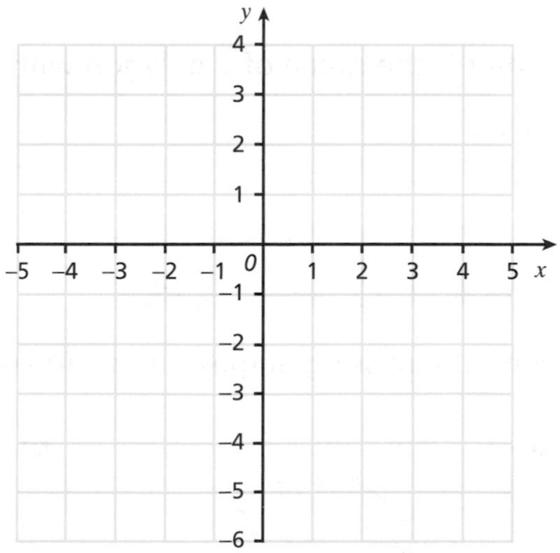

[4]

2 Copy and complete the table using the equation below.

$y = x^2 + 5x + 1$

x	–3	–2	–1	0	1	2	3
y							

[3]

Total Marks .... / 7

Practice Questions

Angles

1 Find the missing angles in the following diagrams.

a) 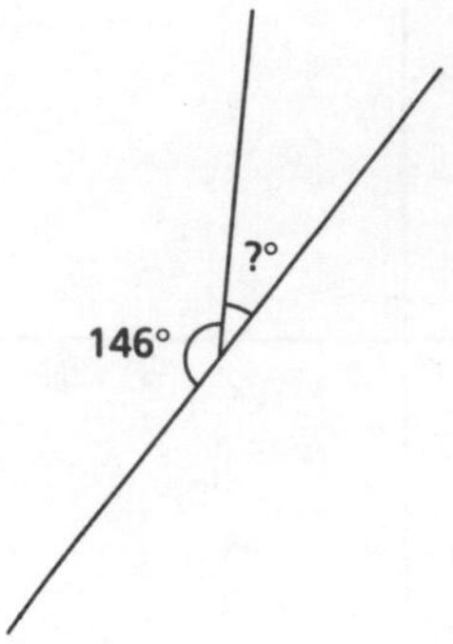

b) 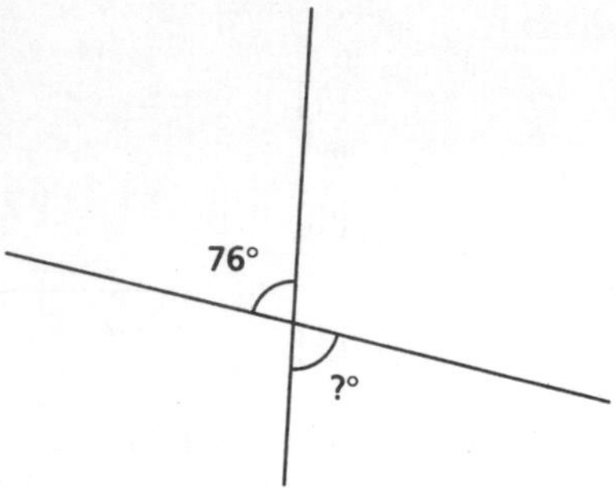

c) 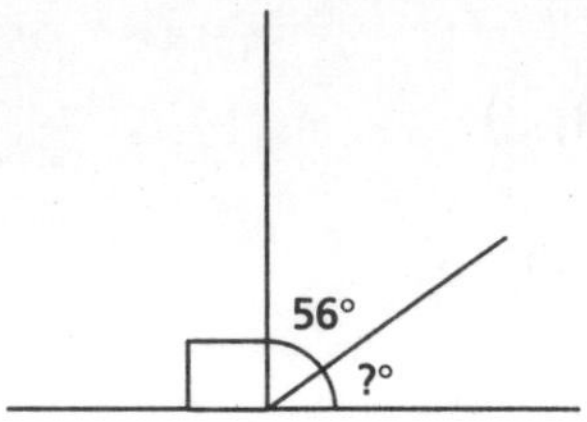

[3]

2 Find the missing angle in this isosceles triangle.

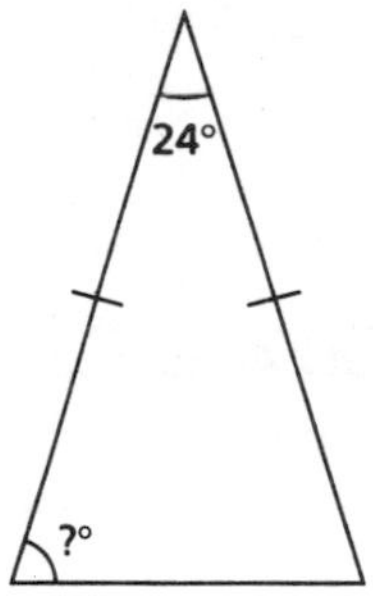

[2]

3 What is the name of a polygon with seven sides? [1]

Total Marks .... / 6

1 Find the missing angles x and y indicated in each diagram.

a) 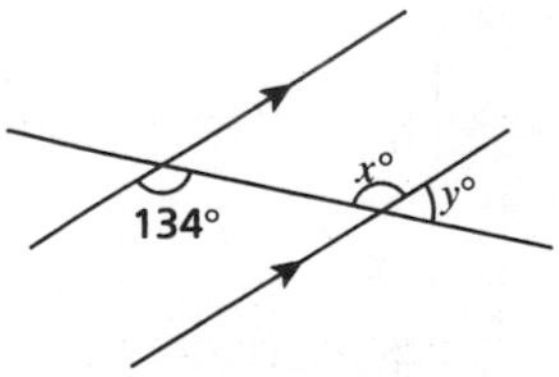

b) 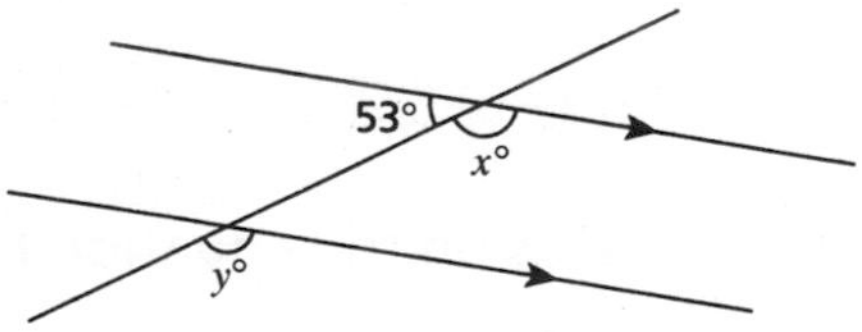

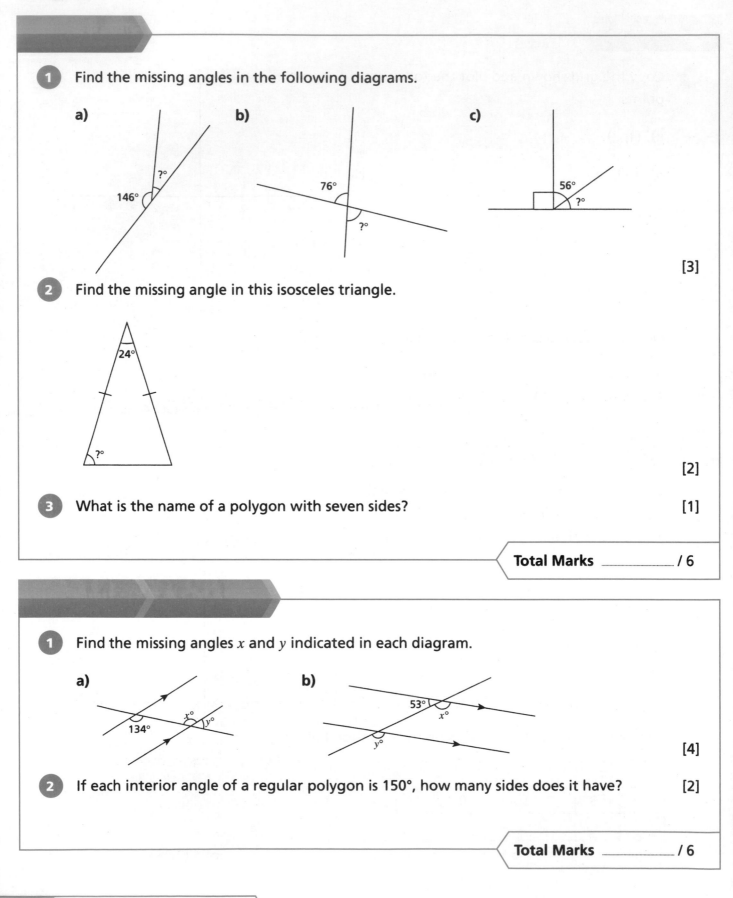

[4]

2 If each interior angle of a regular polygon is 150°, how many sides does it have? [2]

Total Marks .... / 6

Probability

1. If today is Sunday, what is the chance that tomorrow is Tuesday? [1]
2. An eight-sided spinner is numbered 0, 0, 0, 1, 1, 1, 1, 2.

   a) What is the probability of getting a 0 on one spin? [1]

   b) What is the probability of **not** getting a 2? [2]

Total Marks .... / 4

1. Phil does an experiment dropping a paper cup a number of times. The probability of it landing upside-down is 0.65

   What is the probability of it **not** landing upside-down? [2]
2. A dice has been rolled 50 times and recorded in the frequency table below.

   a) Copy and complete the table. Give the probabilities as fractions.

Number	Frequency	Estimated probability
1	5	
2	8	
3	7	
4	7	
5	8	
6	15	
	Total 50	1

[3]

   b) Use the results in your table to work out the estimated probability (as fractions) of getting:

   i) the number 6 **ii)** an odd number **iii)** a number bigger than 4 [3]

Total Marks .... / 8

Fractions, Decimals and Percentages 1

You must be able to:

- Convert between a fraction, decimal and percentage
- Calculate a fraction of a quantity
- Calculate a percentage of a quantity
- Compare quantities using percentages.

Different Ways of Saying the Same Thing

- The table below shows how **fractions, decimals** and **percentages** are used to represent the same thing:

Picture	Fraction	Decimal	Percentage
$\frac{1}{4}$	$\frac{1}{4}$	0.25	25%
$\frac{1}{2}$	$\frac{1}{2}$	0.5	50%
$\frac{3}{4}$	$\frac{3}{4}$	0.75	75%

Key Point

Percent means **'out of 100'**.

Converting Fractions to Decimals to Percentages

- For conversions you do not know automatically, use the rules below.

Fraction $\frac{1}{5}$

↓ **Divide top by bottom** (1 ÷ 5) ↓ ↑ **Put over 100 and cancel down** $\left(\frac{20}{100}\right)$ ↑

Decimal 0.2

↓ **× 100** (0.2 × 100) ↓ ↑ **÷ 100** (20 ÷ 100) ↑

Percentage 20%

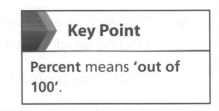

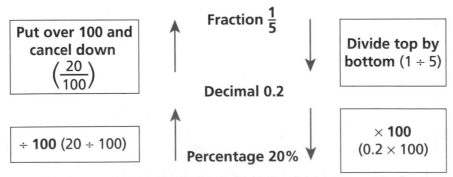

Fractions of a Quantity

- To find a fraction of a **quantity** without a calculator, divide by the **denominator** and multiply by the **numerator**.

Example

Find $\frac{2}{3}$ of £120.

$= 120 \div 3 \times 2 = £80$

- To find a fraction of a quantity with a calculator, use the fraction button $\boxed{a^{b}/c}$. Calculators can differ though, so find out how yours deals with fractions.

Example
Find $\frac{2}{3}$ of £120.

$= 2 \boxed{a^{b}/c} 3 \times 120 = £80$

Percentages of a Quantity

- To find a percentage of a quantity without a calculator:

Example
Find 20% of \$60.

10% of \$60 = 60 ÷ 10 = \$6
20% = \$6 × 2 = \$12

20% is two lots of 10%

- To find a percentage of a quantity with a calculator:

Example
Find 20% of \$60.

= 20% × \$60
= 20 ÷ 100 × 60 = \$12

Or 20% = 0.2
so 20% of \$60 is
0.2 × \$60 = \$12

Change the percentage to a decimal.

Example
Find 120% of \$60.

= 120% × \$60
= 120 ÷ 100 × 60 = \$72

Or 120% = 1.2
so 120% of \$60 is
1.2 × \$60 = \$72

Remember: as 120% > 100%, your answer will be bigger than \$60.

- To compare two quantities using percentages:

Example
Which is bigger, 30% of £600 or 25% of £700?

10% of £600 = 600 ÷ 10 = £60
30% of £600 = 60 × 3 = £180

50% of £700 = 700 ÷ 2 = £350
25% of £700 = 350 ÷ 2 = £175

So 30% of £600 is bigger, by £5.

Key Point

Useful percentages to know:

50% → ÷ 2
10% → ÷ 10
1% → ÷ 100
'of' means '×'

Key Words

fraction
decimal
percentage
quantity
denominator
numerator

Quick Test

1. Change $\frac{7}{20}$ to a decimal and a percentage.
2. Change 36% to a fraction in its lowest terms.
3. Work out $\frac{2}{5}$ of \$70.
4. Find 35% of \$140.

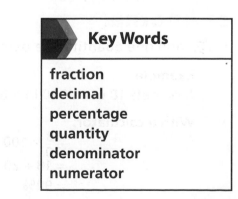

Fractions, Decimals and Percentages 2

You must be able to:

- Work out a percentage increase or decrease
- Find one quantity as a percentage of another
- Work out simple interest.

Percentage Increase and Decrease

- To find a percentage increase or decrease, add on or subtract the percentage you have found.

Example
A calculator is priced at £12 but there is a discount of 25%.
Work out the reduced price of the calculator.

25% of £12
= 25% × £12
= 25 ÷ 100 × 12 = £3

Reduced price = £12 – £3 = £9

£3 is the discount so **'take it away'** to get the final answer.

A one-step way:
A reduction of 25% means you are left with 75%, and 75% = 0.75
0.75 × £12 = £9

Example
A laptop computer costs £350 plus tax at 20%.
Work out the actual cost of the laptop.

20% of £350
= 20% × £350
= 20 ÷ 100 × 350 = £70

£70 is the tax so **'add it on'** to get the final answer.

Actual cost = £350 + £70 = £420
A one-step way:
An increase of 20% means you pay 120%, and 120% = 1.2
1.2 × £350 = £420

Finding One Quantity as a Percentage of Another

- To find one quantity as a percentage of another:

Example
Jane gets 18 out of 20 in a test. What percentage is this?

With a calculator:

$\frac{18}{20} \times 100$
= 18 ÷ 20 × 100
= 90%

Change the fraction to a decimal.

Without a calculator:

$$\frac{18}{20} \overset{\times 5}{=} \frac{90}{100} = 90\%$$

Make the fraction 'out of' 100.

Simple Interest

- Find the **interest** for one year then multiply by the number of years.

Example

Peter puts £200 into a savings account. He gets 5% simple interest per year.

How much does he have in his account after two years?

10% of £200
= 200 ÷ 10
= £20

5% = 20 ÷ 2 = £10

This is the interest for one year.

After two years Peter receives £10 × 2 = £20

Peter has £200 + £20 = £220 in his account.

Key Point

Interest is **not** added on at the end of each year.

Example

Sarah borrows £2000 from her bank. She has to pay back the loan with additional interest at 8% per year.

After three years, how much will Sarah have to pay back?

8% of £2000
= 8% × £2000
= 8 ÷ 100 × 2000 = £160

This is the interest for one year.

After three years the interest will be £160 × 3 = £480

So Sarah will have to pay back £2000 + £480 = £2480

Quick Test

1. A television costing £450 is reduced by 10%. What is its sale price?
2. A house costing £80 000 increases in value by 15%. What is the cost of the house now?
3. Anne gets $\frac{21}{25}$ in a test. What percentage is this?

Key Words

increase
decrease
interest

Algebra

Equations 1

You must be able to:

- Find an unknown number
- Solve a simple equation
- Solve an equation with unknowns on both sides
- Apply the inverse of an operation.

Finding Unknown Numbers

- The unknown number is usually given as a letter or symbol.
- Applying the **inverse** operations helps you to find the unknown number.

Example

$n + 4 = 13$ or ■ $+ 4 = 13$

Take 4 away from 13.

n or ■ $= 13 - 4$

n or ■ $= 9$ (check: $9 + 4 = 13$)

Both ways mean **'something + 4 = 13'**.

– 4 is the 'inverse' or 'opposite' of + 4.

Example

$x - 4 = 13$ or ● $- 4 = 13$

Add 4 to 13.

x or ● $= 13 + 4$

x or ● $= 17$ (check: $17 - 4 = 13$)

Both ways mean **'something – 4 = 13'**.

+ 4 is the 'inverse' or 'opposite' of – 4.

- The same idea applies to multiplying and dividing.

Example

$3n = 12$

$n = 12 \div 3$

$n = 4$ (check: $3 \times 4 = 12$)

3 × something = 12

÷ 3 is the inverse of × 3

Example

$\frac{n}{3} = 4$

$n = 4 \times 3$

$n = 12$ (check: $12 \div 3 = 4$)

Something ÷ 3 = 4

× 3 is the inverse of ÷ 3

- Now you can apply inverse operations to more difficult **equations**.

Key Point

Operation	Inverse
+	–
–	+
×	÷
÷	×

Solving Equations

- Remember to think of the letter as 'something'.

Example

Solve the equation $2y + 3 = 15$

This simply means 'something' + 3 = 15

'Something' must be 12.

So $2y = 12$ means 2 × 'something' = 12

'Something' must be 6.

So $y = 6$

Now look with the **inverses**:

$$2y + 3 = 15$$

$(-3) \quad 2y = 12$

$(\div 2) \quad y = 6$

Key Point

Remember to do the same to both sides of the equation.

Equations with Unknowns on Both Sides

- An equation may have an unknown number on both sides of the equals sign.

Example

Solve the equation $5x - 2 = 3x + 5$

$5x - 2 = 3x + 5$

Do the same to both sides.

$(-3x) \quad 2x - 2 = 5$

Get rid of all the xs on one side of the equation.

$(+2) \quad 2x = 7$

Now do the inverses.

$(\div 2) \quad x = 3.5$

Example

Solve the equation $3x + 5 = 5x - 4$

$3x + 5 = 5x - 4$

$(-3x) \quad 5 = 2x - 4$

$(+4) \quad 9 = 2x$

$(\div 2) \quad 4.5 = x$

This is the same as $x = 4.5$

Quick Test

1. What is the inverse of × 6?
2. If ● – 5 = 7, what is the value of ●?
3. If $6n = 30$, what is the value of n?
4. Solve the equation $3y - 2 = 13$
5. Solve the equation $3x + 7 = 2x - 2$

Key Words

inverse
equation
solve

Algebra

Equations 2

You must be able to:

- Solve more complex equations with fractions or negative numbers
- Set up and solve an equation
- Apply the inverse of an operation
- Use negative numbers.

Solving More Complex Equations

- An equation may include a **negative** of the unknown number.
- The unknown number may be part of a fraction or inside **brackets**.

Example

Solve the equation $3x + 1 = 11 - 2x$

$$3x + 1 = 11 - 2x$$

$(+ 2x)$ $\quad 5x + 1 = 11$

$(- 1)$ $\quad 5x = 10$

$(\div 5)$ $\quad x = 2$

Be careful. This equation has some **negative** values of x. Adding $2x$ to both sides gets rid of the negative xs.

Example

Solve the equation $\frac{3x+1}{2} = 8$

$$\frac{3x+1}{2} = 8$$

$(\times 2)$ $\quad 3x + 1 = 16$

$(- 1)$ $\quad 3x = 15$

$(\div 3)$ $\quad x = 5$

Here the × 2 cancels out the ÷ 2 on the left-hand side of the equation.

Example

Solve the equation $3(2x - 1) = 4(x + 2)$

Multiply out the brackets first then solve in the usual way:

$$3(2x - 1) = 4(x + 2)$$

$$6x - 3 = 4x + 8$$

$(- 4x)$ $\quad 2x - 3 = 8$

$(+ 3)$ $\quad 2x = 11$

$(\div 2)$ $\quad x = 5.5$

Key Point

Remember to multiply everything inside the bracket by the number outside the bracket.

Setting Up and Solving Equations

- You will have to follow a set of instructions in a given order.
- Usually you only have to ×, ÷, +, – or square.
- If you have to multiply 'everything' then remember to use brackets.

Example

A number is **doubled**, then 5 is added to the total and the result is 11.

What was the original number?

The words	The algebra
A number	n
doubled	$2n$
add 5	$2n + 5$
the result is 11	$2n + 5 = 11$

You can now solve this in the usual way to find the original number equals 3.

Example

Three boys were paid £10 per hour plus a tip of £6 to wash some cars. They shared the money and each got £12. How many hours did they wash cars for?

£10 × number of hours

£6 tip

Set up an equation: $\frac{10x + 6}{3} = 12$

each got

shared by 3 boys

Now solve the equation:

(× 3) $10x + 6 = 36$

(– 6) $10x = 30$

(÷ 10) $x = 3$

The boys worked for 3 hours.

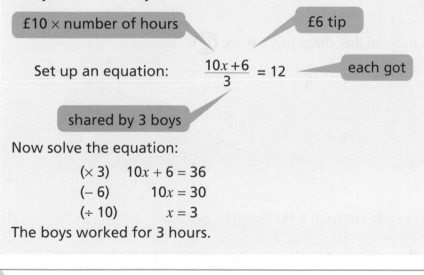

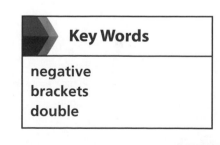

Key Point

Remember to do the same to both sides of the equation.

Quick Test

1. Solve the equation $3x - 4 = 11$
2. Solve the equation $2x + 3 = 12 - x$
3. Solve the equation $2(x + 2) = 2(3x - 2)$
4. A number is multiplied by 3 and then 8 is subtracted. The result is 25. What is the number?
5. A chicken is roasted for 60 minutes for every kilogram, plus an extra 20 minutes. If the chicken took 140 minutes to cook, how heavy was it?

Key Words

negative
brackets
double

Review Questions

Angles

1 Find the missing angles.

a) 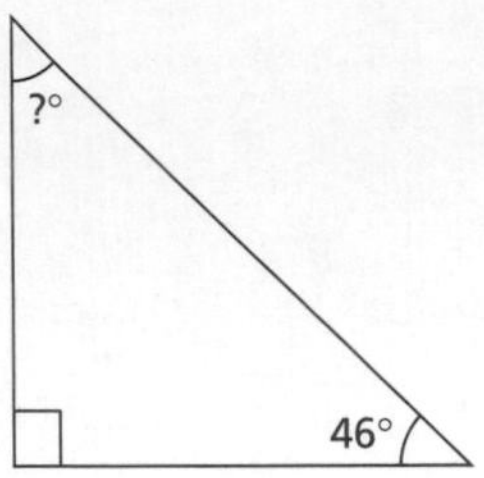

b) 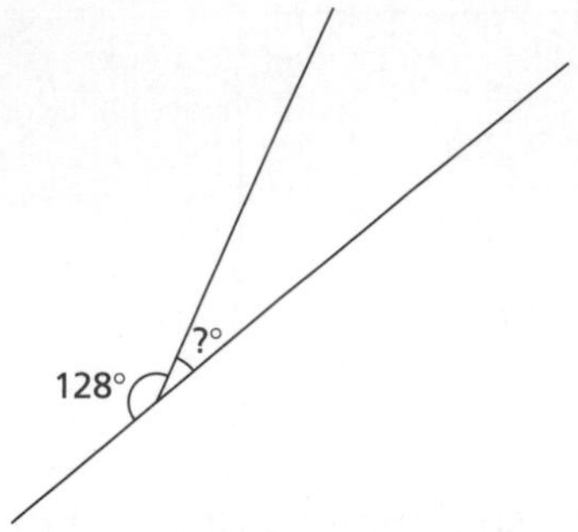

c) 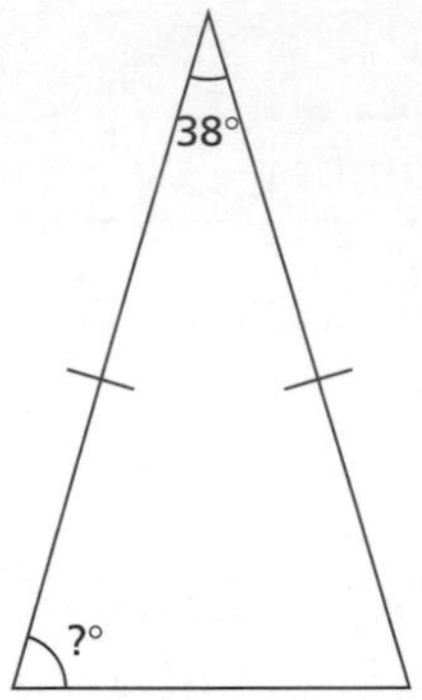

[3]

Total Marks .... / 3

1 Find the missing angles in the parallel line diagrams below.

a) 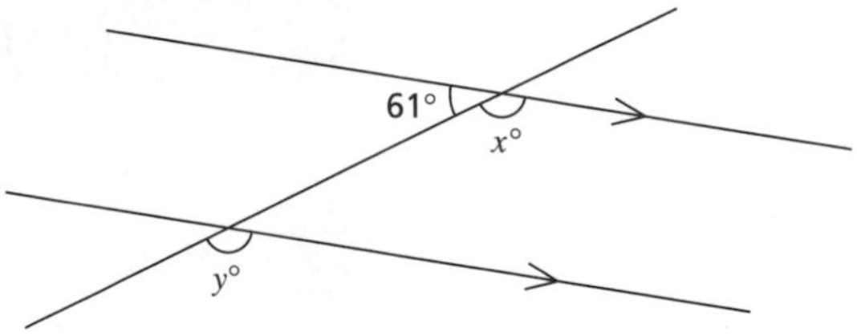

b) 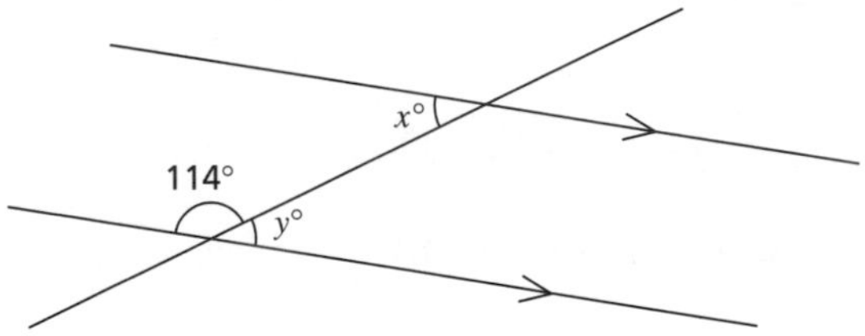

[4]

2 What do the total interior angles add up to in a nonagon? [1]

3 If each interior angle of a regular polygon equals 160°, how many sides does it have?

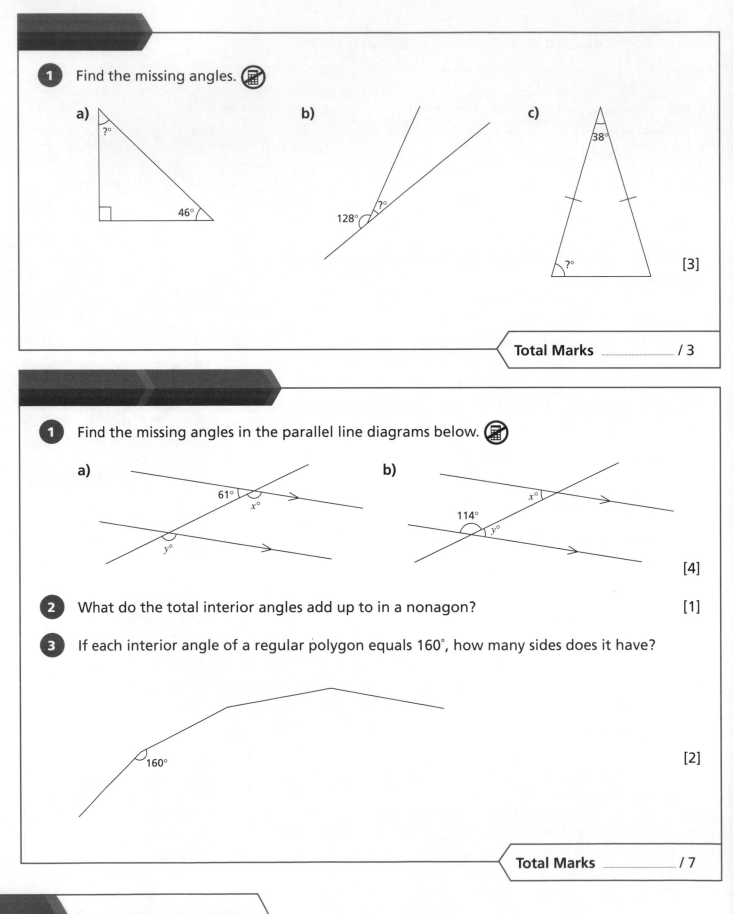

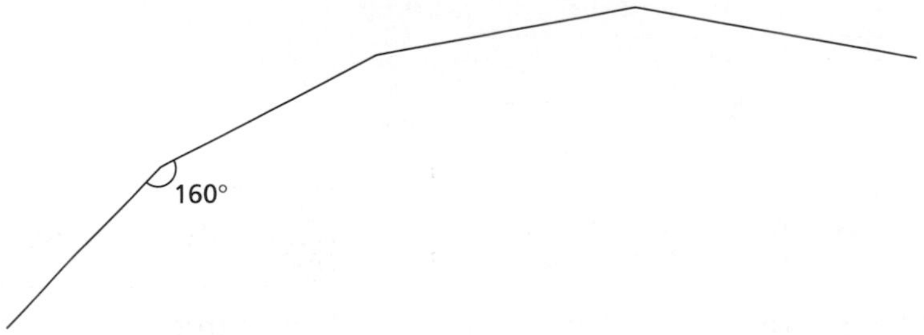

[2]

Total Marks .... / 7

Probability

1. Copy the probability scale and complete the boxes with suitable probability words.

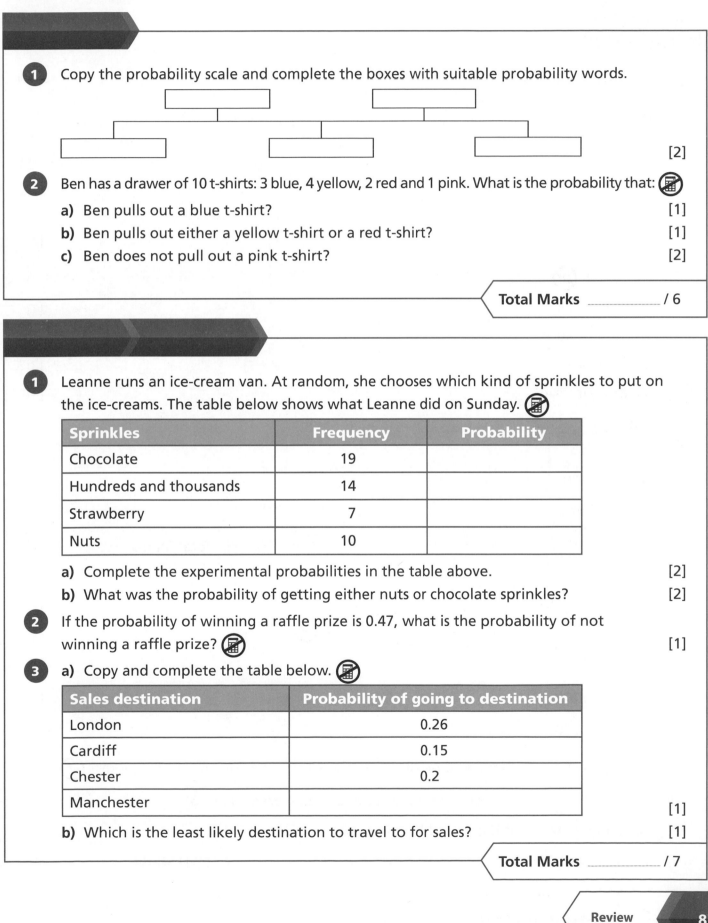

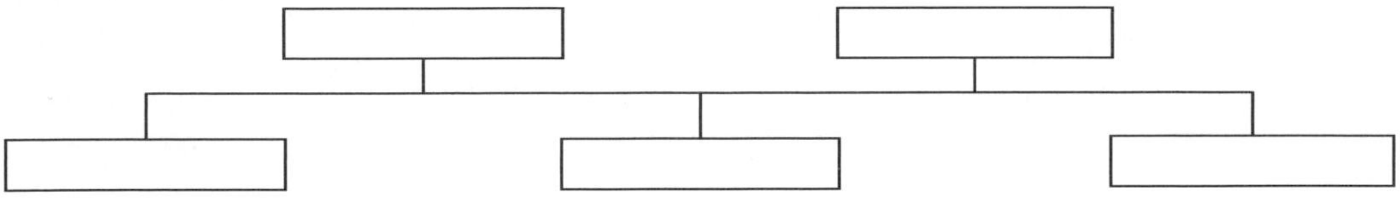

[2]

2. Ben has a drawer of 10 t-shirts: 3 blue, 4 yellow, 2 red and 1 pink. What is the probability that:

   a) Ben pulls out a blue t-shirt? [1]

   b) Ben pulls out either a yellow t-shirt or a red t-shirt? [1]

   c) Ben does not pull out a pink t-shirt? [2]

Total Marks .... / 6

1. Leanne runs an ice-cream van. At random, she chooses which kind of sprinkles to put on the ice-creams. The table below shows what Leanne did on Sunday.

Sprinkles	Frequency	Probability
Chocolate	19	
Hundreds and thousands	14	
Strawberry	7	
Nuts	10	

   a) Complete the experimental probabilities in the table above. [2]

   b) What was the probability of getting either nuts or chocolate sprinkles? [2]

2. If the probability of winning a raffle prize is 0.47, what is the probability of not winning a raffle prize? [1]

3. **a)** Copy and complete the table below.

Sales destination	Probability of going to destination
London	0.26
Cardiff	0.15
Chester	0.2
Manchester	

[1]

   b) Which is the least likely destination to travel to for sales? [1]

Total Marks .... / 7

Practice Questions

Fractions, Decimals and Percentages

1. Copy and complete the following table.

Fraction	Decimal	Percentage
$\frac{7}{10}$		
		55
	0.32	
$\frac{3}{100}$		

[4]

2. Work out:

a) 50% of £32 [1]

b) 10% of 80cm [1]

c) 15% of 160m [2]

d) 25% of £104 [1]

3. Use the a^{b}/c button on your calculator to work out:

a) $\frac{1}{5}$ of £85 [1]

b) $\frac{2}{3}$ of £120 [1]

c) $\frac{5}{7}$ of 21m [1]

Total Marks .... / 12

1. Work out the following, showing your working.

a) $\frac{2}{3}$ of £15 [3]

b) $\frac{3}{7}$ of £210 [3]

c) $\frac{4}{5}$ of £6000 [3]

PS 2. A jacket costing £75 is reduced by 20% in a sale. What is the sale price of the jacket? [3]

PS 3. In a Maths test, Karim scored 16 out of 20 and John scored 15 out of 20.

What percentage did Karim and John each get? [4]

Total Marks .... / 16

Equations

PS **1** Use inverses to work out the unknown number.

a) $\triangle - 7 = 3$ [1]

b) $18 + \triangle = 23$ [1]

c) $\frac{n}{4} = 3$ [1]

d) $5y = 35$ [1]

PS **2** Solve these equations.

a) $3n + 1 = 13$ [2]

b) $2x - 5 = 3$ [2]

c) $5y + 1 = 11$ [2]

Total Marks .... / 10

PS **1** Solve these equations.

a) $3x + 1 = x + 7$ [2]

b) $2(2x - 3) = x - 3$ [2]

c) $6(x + 1) = 2(x + 13)$ [2]

d) $\frac{3x+5}{4} = 5$ [2]

PS **2** A number is multiplied by 3, then 2 is added to the total and the result is 11.

Set up and solve an equation to find the original number. [2]

PS **3** Mark adds together his age and the age of his sister. He gets a total of 28.

If Mark is 16, how old is his sister? Set up and solve an equation to find your answer. [2]

Total Marks .... / 12

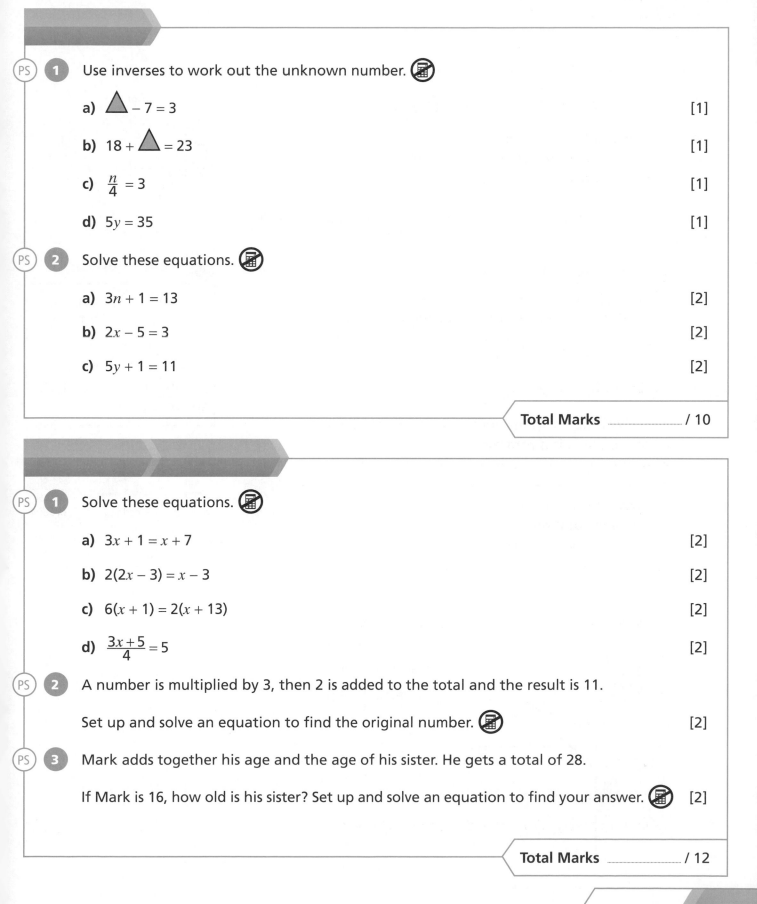

Symmetry and Enlargement 1

You must be able to:

- Reflect a shape
- Translate a shape
- Find the order of rotational symmetry
- Rotate a shape
- Enlarge a shape.

Reflection and Reflectional Symmetry

- Reflect each point one at a time.
- Use a line that is **perpendicular** to the mirror line.
- Make sure the **reflection** is the same distance from the mirror line as the original shape.
- A shape has reflectional symmetry if you can draw a mirror line through it.

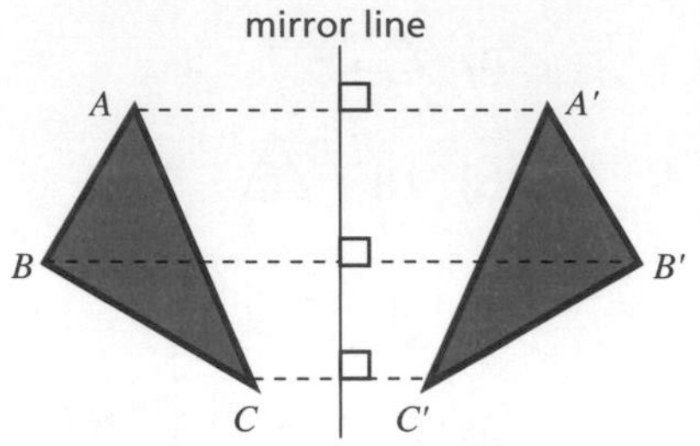

Key Point

Perpendicular means at right angles (90°) to.

Check the position of your reflection by placing a mirror along the mirror line.

Translation

- Translation shifts a shape left/right (x) and/or up/down (y).
- The shift is given as a vector $\begin{pmatrix} x \\ y \end{pmatrix}$.

Rotational Symmetry

- A shape has rotational symmetry if it looks exactly like the original shape when it is **rotated**.
- The **order of rotational symmetry** is the number of ways the shape looks the same.
- To rotate a shape you need to know:
  - The **centre of rotation**
  - The direction it will rotate
  - The number of degrees to rotate it.

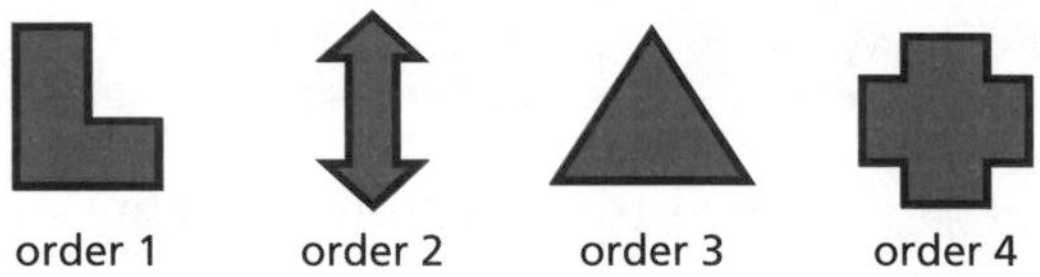

Example

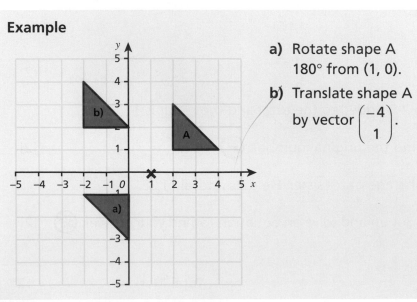

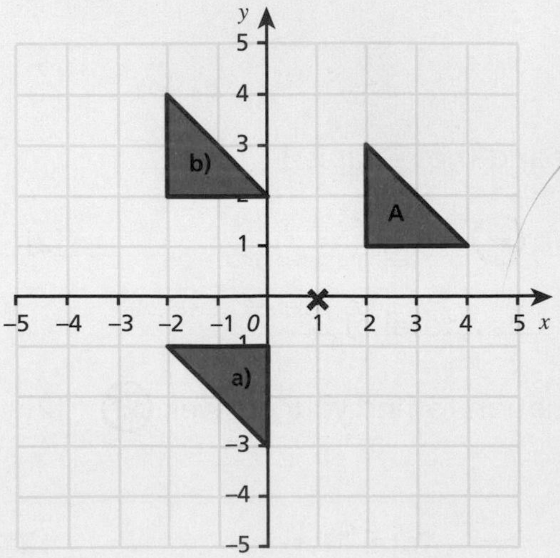

a) Rotate shape A 180° from (1, 0).

b) Translate shape A by vector $\begin{pmatrix} -4 \\ 1 \end{pmatrix}$.

Enlargement

- To draw an **enlargement** you need to know two things:
  - How much bigger/smaller to make the shape. This is called the **scale factor**.
  - Where you will enlarge the shape from. This is called the **centre of enlargement**.
- Remember that enlarged shapes are **similar** (the same shape but a different size).

Example

Enlarge shape A by a scale factor of 2 from the point (3, 4).

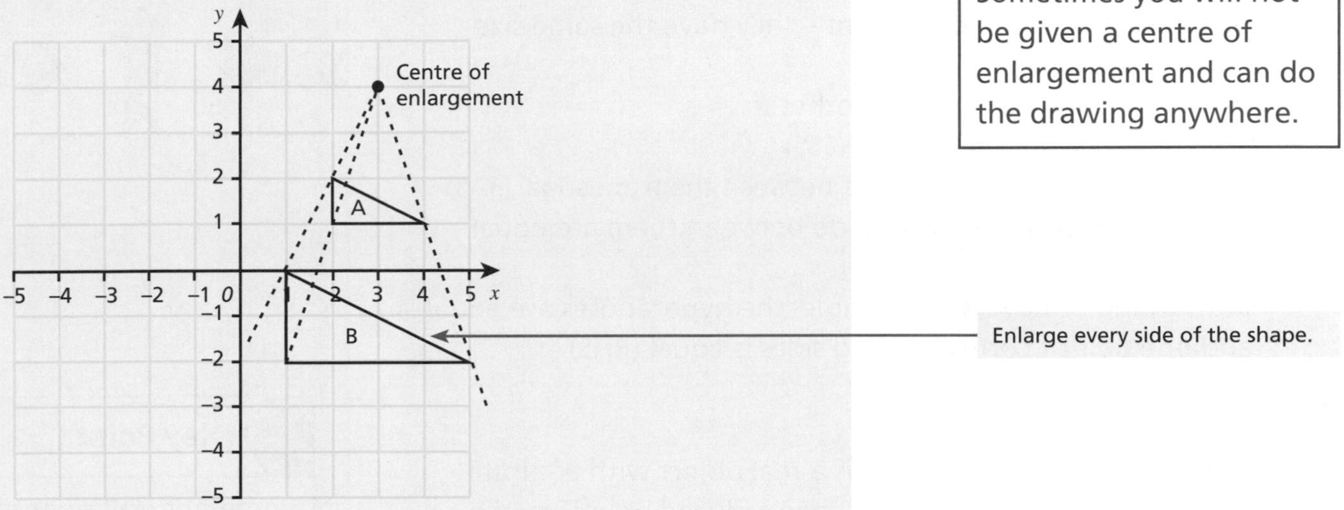

- Use **rays** to check the position of your enlargement. They will touch corresponding corners of the shape.

Key Point

If the scale factor is more than 1, the shape will be bigger.

If the scale factor is less than one, e.g. $\frac{1}{2}$, the shape will be smaller.

Sometimes you will not be given a centre of enlargement and can do the drawing anywhere.

Quick Test

1. What is the order of rotational symmetry of this shape?

2. a) Reflect this shape across the mirror line.
   b) Rotate the shape 180° about the corner A.

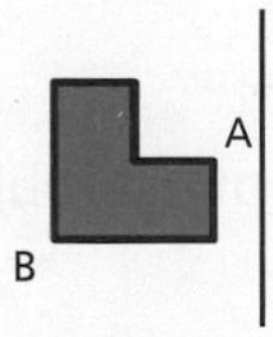

   c) Rotate the shape 90° clockwise about the corner B.

3. Enlarge this shape by a scale factor of 3.

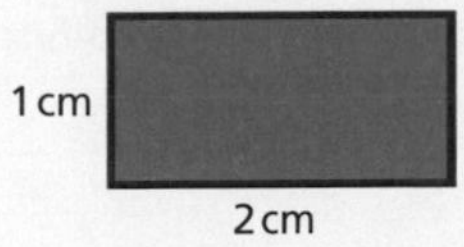

Key Words

perpendicular
reflection
rotation
centre of rotation
enlargement
scale factor
centre of enlargement
similar
ray

Symmetry and Enlargement 2

You must be able to:
- Recognise congruent shapes
- Interpret a scale drawing
- Work out missing sides in similar shapes
- Convert between units of measure.

Congruence

- Congruence simply means shapes that are exactly the same.
- These arrow shapes are **congruent** – they have the same size and the same shape.
- Triangles are congruent to each other if:
  - three pairs of sides are equal (SSS)
  - two pairs of sides and the angle between them are equal (SAS)
  - two pairs of angles and the side between them are equal (ASA)
  - both triangles have a right angle, the hypotenuses are equal and one pair of corresponding sides is equal (RHS).

Same size Same shape

Scale Drawings

- A scale drawing is one that shows a real object with accurate dimensions, except they have all been reduced or enlarged by a certain amount (called the **scale**).
- Similar shapes are enlarged by the same scale factor, but the angles stay the same.

Key Point

Remember, **all** sides must be multiplied by the same number.

Example
A scale of 1 : 10 means in the real world the object would be 10 times bigger than in the drawing. These horses are similar – same shape, different size.

- We use scale drawings to make copies of real objects.

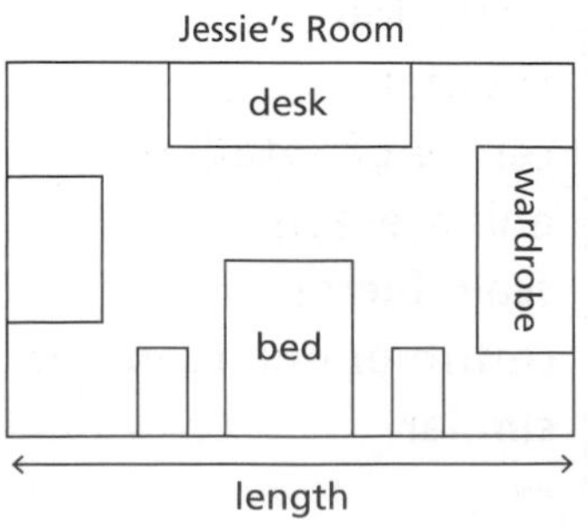

Scale: 1 inch = 3 feet

Here you would have to measure each side in inches then **multiply by 3** to get the real length in feet.

Shape and Ratio

- You can use **ratios** to work out the 'real' size of an object. The scale is given as a ratio with the smaller **unit** first.

Example

Estimate the height of this house using the scale of 1 : 160

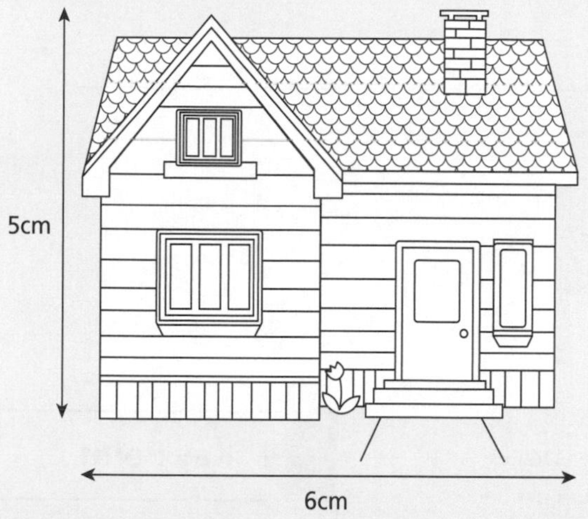

For every 1cm you measure in the picture, multiply by 160 to get the real size. Then convert to metres.

Here, the height of the house is 5cm so in real life the house is actually:

$5 \times 160 = 800\text{cm} = 8\text{m}$

Remember: 100cm = 1m

The width of the house is 6cm so in real life the house is actually:

$6 \times 160 = 960\text{cm} = 9\text{m } 60\text{cm}$

Key Point

Scales are given as a ratio, usually 1 : n where n is what you multiply by.

Quick Test

1. Draw three congruent shapes.
2. Which shape is congruent to shape A?

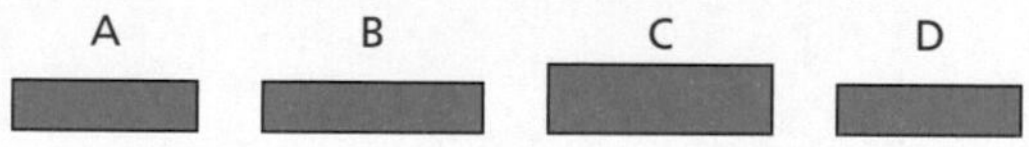

3. Estimate the length, in metres, of the boat. Scale: 1 : 80

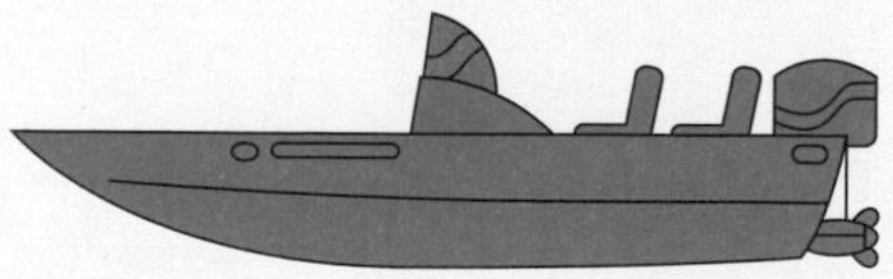

4. Which of these shapes are similar?

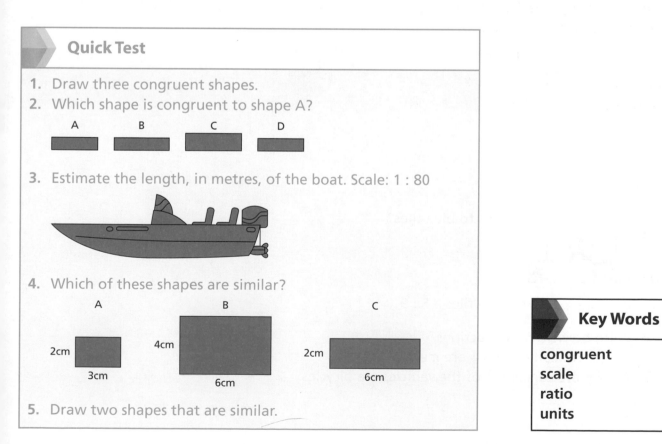

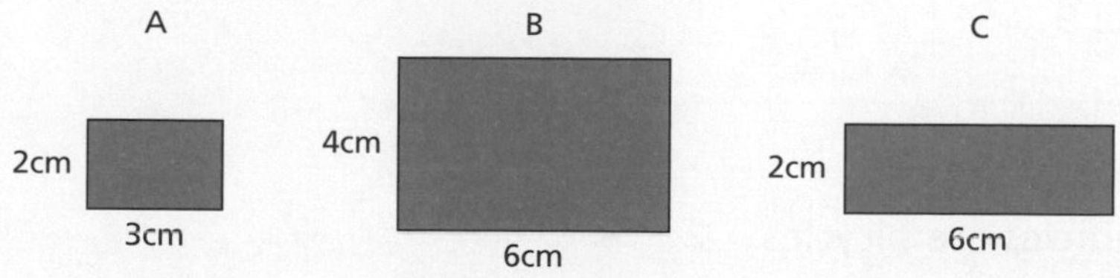

5. Draw two shapes that are similar.

Key Words

congruent
scale
ratio
units

Ratio and Proportion 1

You must be able to:

- Understand what ratio means
- Simplify a ratio
- Multiply and divide by whole numbers.

Introduction to Ratios

- Ratio is a way of showing the relationship between two numbers.
- Ratios can be used to compare costs, weights and sizes.

Example

On the deck of a boat there are 2 women and 1 man. There are also 5 cars and 2 bicycles.

The ratio of men to women is 1 to 2, written 1 : 2

The ratio of women to men is 2 to 1, written 2 : 1

The ratio of cars to bicycles is 5 to 2, written 5 : 2

The ratio of bicycles to cars is 2 to 5, written 2 : 5

Key Point

'to' is replaced with ':'

Example

A recipe for making pastry uses 4oz flour and 2oz butter.

The ratio of flour to butter is 4 to 2, written 4 : 2

The ratio of butter to flour is 2 to 4, written 2 : 4

Example

What is the ratio of black tiles to blue tiles?

The ratio of black tiles to blue tiles is 5 : 9

- Ratios can also be written as fractions.
  For example, $\frac{2}{3}$ are women and $\frac{1}{3}$ are men.
  $\frac{5}{7}$ of the vehicles are cars and $\frac{2}{7}$ of the vehicles are bicycles.

Simplifying Ratios

- The following ratios are **equivalent**. The relationship between each pair of numbers is the same:

10 : 20
↓
3 : 6
↓
2 : 4
↓
1 : 2   This is a **simpler** way of writing the ratio 10 : 20

- You can **simplify** a ratio if you can divide by a common **factor**.
- When a ratio cannot be simplified, it is said to be in its **lowest terms**.

Key Point

When there are no more common factors, the ratio is in its lowest terms.

Example

Simplify the ratio 30 : 100

30 ÷ 10 → 3 : 10 ← 100 ÷ 10

Divide both numbers by 10.

Example

Write 40 cents to $1 as a ratio in its lowest terms.

First get the units the same: 40 cents to 100 cents written

40 : 100

Now simplify (÷ 10) 4 : 10

and again (÷ 2) 2 : 5

This is now in its lowest terms.

Example

The angles of a triangle are 20°, 60° and 100°.

What is the ratio of the angles in its lowest terms?

20° : 60° : 100°

(÷ 20) 1 : 3 : 5

This is now in its lowest terms.

Quick Test

1. Look at this pattern of grey and green tiles:

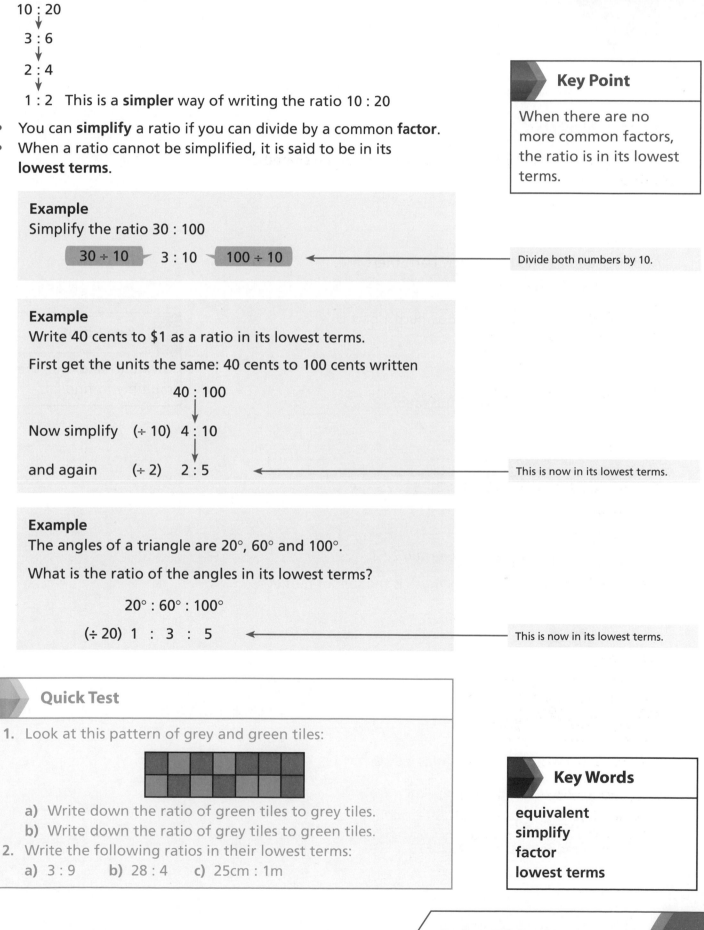

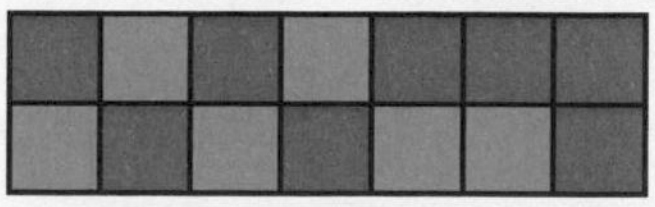

   a) Write down the ratio of green tiles to grey tiles.
   b) Write down the ratio of grey tiles to green tiles.
2. Write the following ratios in their lowest terms:
   a) 3 : 9   b) 28 : 4   c) 25cm : 1m

Key Words

equivalent
simplify
factor
lowest terms

Ratio and Proportion 2

You must be able to:

- Share in a given ratio
- Multiply and divide by whole numbers
- Work out proportional amounts
- Use the unitary method.

Sharing Ratios

- Sharing ratios are used when a total amount is to be **shared** or **divided** into a given ratio.

Example
Share £200 in the ratio 5 : 3

Add the ratio to find how many parts there are.

$5 + 3 = 8$ parts

Divide £200 by 8 to find out how much 1 part is.

$200 \div 8 = 25$

1 part is £25

Now multiply by each part of the ratio.

$5 \times £25 = £125$

$3 \times £25 = £75$

£200 shared in the ratio 5 : 3 is £125 : £75

Key Point

Divide to find one, then multiply to find all.

Example
A sum of money is shared in the ratio 2 : 3

If the smaller share is £30, how much is the sum of money?

2 parts = £30

so 1 part = $£30 \div 2 = £15$

3 parts = $£15 \times 3 = £45$

The sum of money = $£30 + £45 = £75$

Direct Proportion

- Two quantities are in **direct proportion** if their ratios stay the same as the quantities get larger or smaller.

Example
If the ratio of teachers to students in one class is 1 : 30, then three classes will need 3 : 90

Using the Unitary Method

- Using the unitary method, find the value of **one unit** of the quantity before working out the required amount.

Example
Five loaves of bread cost £4.25. How much will three loaves cost?

One loaf costs £4.25 ÷ 5 = 85 pence

Three loaves will cost 85 pence × 3 = £2.55

Remember: divide to find one, then multiply to find all.

Example
This recipe for making apple pie serves four people:

200g flour

200g butter

50g sugar

8 large apples

Change these amounts so there are enough ingredients for 10 people.

Divide to find one, then multiply to find all.

200g flour ÷ 4 × 10 = 500g flour

200g butter ÷ 4 × 10 = 500g butter

50g sugar ÷ 4 × 10 = 125g sugar

8 large apples ÷ 4 × 10 = 20 large apples

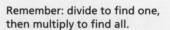

Key Point

All the amounts have increased in proportion (by $2\frac{1}{2}$ times in this example).

Quick Test

1. Share 40 sweets in the ratio 2 : 5 : 1
2. £360 is divided between Sara and John in the ratio 5 : 4
   How much did each person receive?
3. Work out the missing ratio.
   a) 3 : 5 = 12 : ? b) 4 : 5 = ? : 35
4. If six CDs cost £30, how much will eight CDs cost?

Key Words

share
divide
direct proportion

Review Questions

Fractions, Decimals and Percentages

PS **1** Convert:

a) $\frac{3}{20}$ to a percentage. [1]

b) 0.8 to a fraction in its lowest terms. [2]

PS **2** Work out:

a) 50% of £23 [1]

b) 10% of 45cm [1]

c) 25% of 180m [1]

d) 5% of £70 [2]

3 Use the $a^{b}/_{c}$ button on your calculator to work out:

a) $\frac{1}{3}$ of £42 [1]

b) $\frac{3}{4}$ of £320 [1]

c) $\frac{4}{5}$ of £850 [1]

Total Marks .... / 11

FS **1** Jenny receives £5 pocket money every week.

She spends $\frac{1}{2}$ of her money on magazines and $\frac{2}{5}$ on sweets. The rest she saves.

a) How much does Jenny spend on sweets? [2]

b) How much does Jenny save? [2]

PS **2** A coat costing £90 is reduced by 15% in a sale. What is the sale price of the coat? [3]

FS **3** Kim puts £150 into a savings account. She will receive 6% simple interest each year. How much will she have in the bank after four years? [3]

Total Marks .... / 10

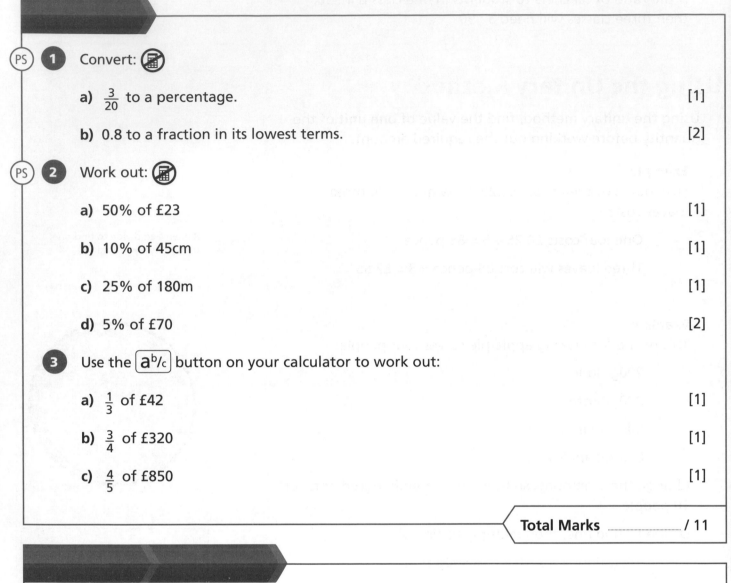

Equations

PS **1** Use inverses to work out the unknown number.

a) 7◯ = 28 [1]

b) $n + 8 = 17$ [1]

c) $\frac{p}{3} = 6$ [1]

d) ◯ − 7 = 12 [1]

PS **2** Solve these equations. Show your working.

a) $4n - 1 = 11$ [2]

b) $5x + 1 = 21$ [2]

c) $3a + 8 = 5$ [2]

Total Marks .... / 10

PS **1** Solve these equations. Show your working.

a) $6x - 5 = 4x + 7$ [2]

b) $5(x + 2) = 2(x - 1)$ [2]

c) $3x - 1 = 4 - 2x$ [2]

FS **2** Five builders are together paid £20 per hour plus a bonus of £150. They share the pay and each get £50.

Set up and solve an equation to find how many hours they worked. [2]

3 A chocolate bar machine holds 56 bars of chocolate.

If 29 are left, how many were sold? Set up and solve an equation to find your answer. [2]

Total Marks .... / 10

Practice Questions

Symmetry and Enlargement

PS **1** Reflect shape A across the dotted mirror lines.

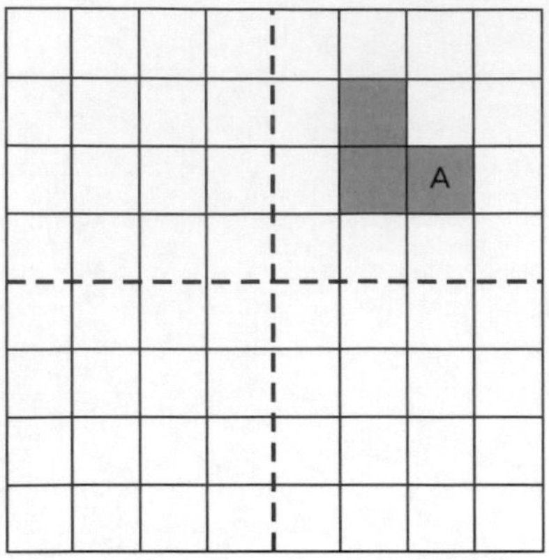

[3]

PS **2** What is the order of rotational symmetry of this shape?

[1]

Total Marks .... / 4

PS **1** **a)** Rotate shape A 90° clockwise about the origin (0, 0). Label the new shape B. [2]

b) Enlarge shape A by a scale factor 3, with a centre of enlargement (3, 4). Label the new shape C. [2]

c) Which shapes are congruent? [1]

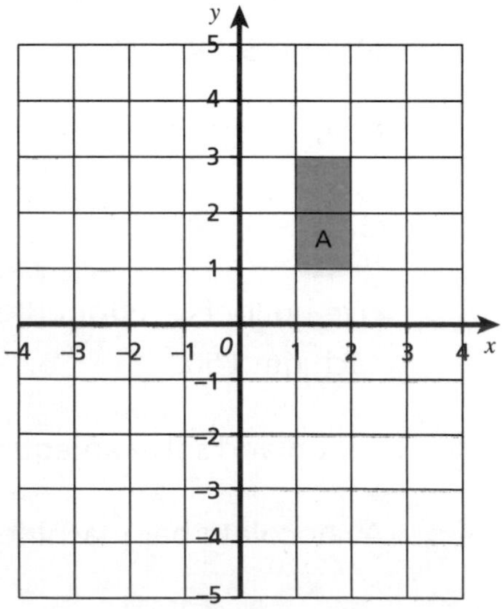

MR **2** A photograph 5cm × 7cm is to be enlarged by a scale factor of 4.

What are the dimensions of the new photograph? [2]

Total Marks .... / 7

Ratio and Proportion

PS **1** What is the ratio of black tiles to white tiles? [1]

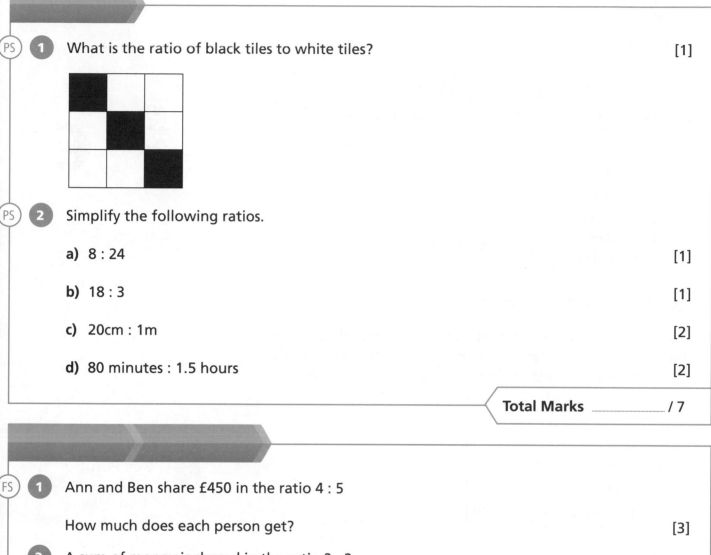

PS **2** Simplify the following ratios.

a) 8 : 24 [1]

b) 18 : 3 [1]

c) 20cm : 1m [2]

d) 80 minutes : 1.5 hours [2]

Total Marks .... / 7

FS **1** Ann and Ben share £450 in the ratio 4 : 5

How much does each person get? [3]

2 A sum of money is shared in the ratio 2 : 3

If the larger share is £27, how much money is there altogether? [3]

PS **3** A recipe for six cupcakes needs 40g of butter and 100g of flour.

How much butter and flour are needed to make 12 cupcakes? [2]

PS **4** A map has a scale of 1 : 50 000.

What is the distance on the ground, in km, if the distance on the map is:

a) 2.5cm? [2]

b) 1.4cm? [2]

Total Marks .... / 12

Real-Life Graphs and Rates 1

You must be able to:

- Read values from a real-life graph
- Draw a real-life graph
- Read a conversion graph
- Draw a conversion graph.

Graphs from the Real World

- **Graphs** from the real world include **conversion** graphs.
- You may be asked to convert between these units:
  - £ and $
  - £ and euros
  - pints and litres
  - mph and km/h
  - miles and km
  - gallons and litres

Reading a Conversion Graph

- To convert from one unit to the other, read straight across to the line, then go straight down until you reach the other axis. If you are converting the other way, go up until you reach the line, then read across.

Example

Convert 30 miles into kilometres.

Draw a line **straight up** from 30 miles until it hits the line.

Go **straight across** to the km axis.

30 miles is **equivalent** to 48km.

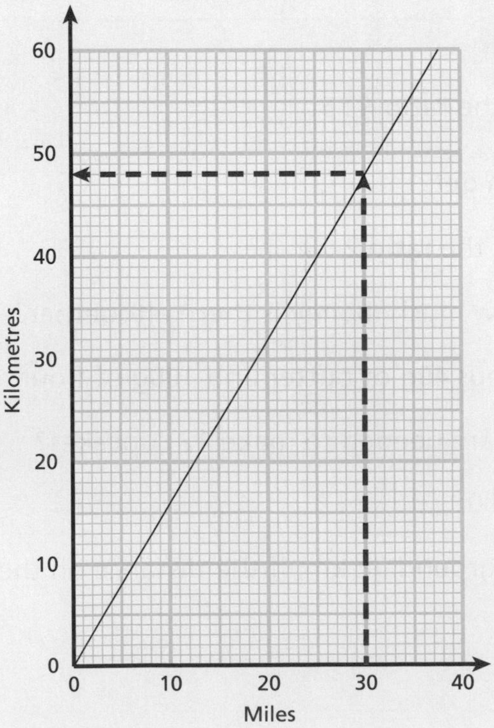

Key Point

To go from kilometres to miles, read straight across to the line, then go straight down until you hit the miles axis.

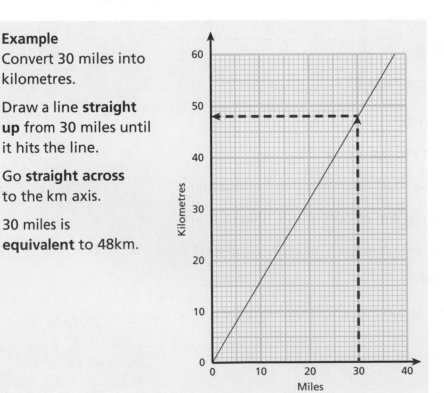

Drawing a Conversion Graph

- To find the points you need to plot, work out a number of equivalent values.
- Plot these values on a graph and join with a straight line.

Example

Jack's company pays him 80 pence for each mile he travels. Use the information to draw a graph of his pay.

First work out how much Jack will be paid for different journeys.

Distance in miles	0	10	20	30
Amount	0	£8	£16	£24

Now plot the points on a graph and join up with a straight line.

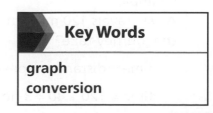

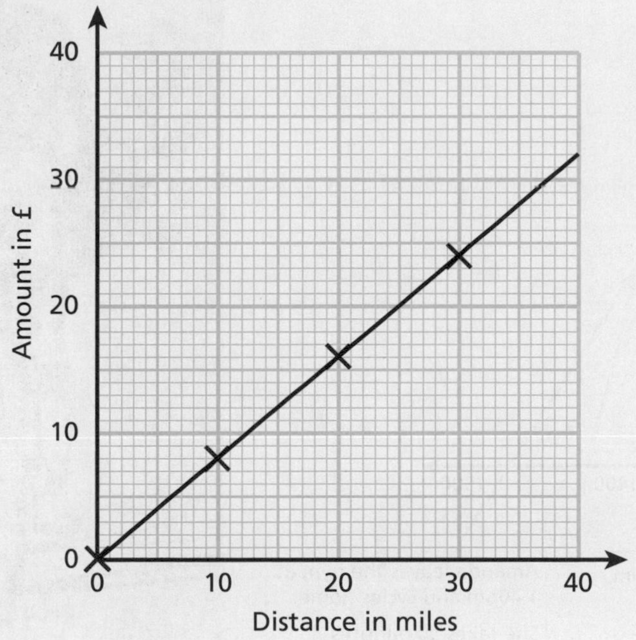

Key Point

Extend your line to the edge of the graph grid.

The graph can now be read to find the pay for different journeys.

Quick Test

Use the conversion graph on page 102 for questions 1 and 2:

1. Find how many kilometres are equivalent to:
   a) 25 miles **b)** 10 miles
2. Find how many miles are equivalent to:
   a) 30km **b)** 40km
3. Sharon charges \$1 for the use of her taxi and 50 cents per mile after that. Work out the cost of a journey that is:
   a) 4 miles **b)** 6 miles long
4. Use the information from question 3 to draw a graph of Sharon's charges.

Key Words

graph
conversion

Real-Life Graphs and Rates 2

You must be able to:
- Read a distance–time graph
- Work out speed, distance and time
- Work out unit prices
- Work out density, mass and volume.

Time Graphs

- Distance–time graphs give information about journeys. Use the horizontal scale for time and the vertical scale for **distance**.
- Distance–time graphs are also used to calculate **speed**.

Example

Amanda cycles to the gym and back every Sunday. The graph below shows Amanda's journey.

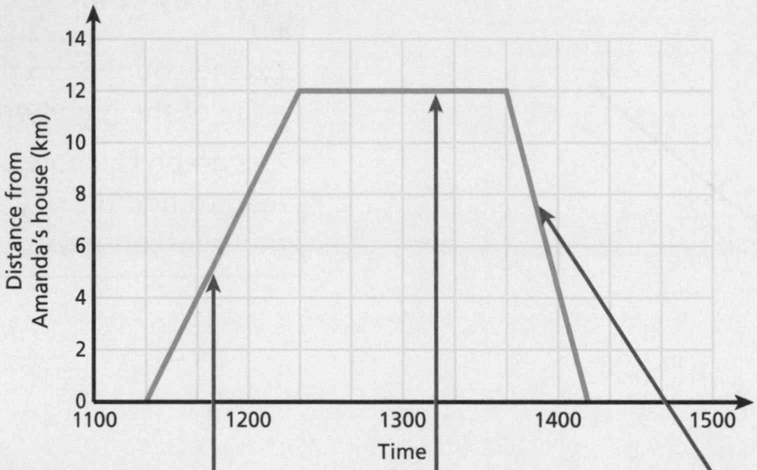

Amanda leaves home at 11.20am and cycles 12km in one hour.

Her speed is 12km/h.

Amanda arrives at the gym at 12.20pm and stays for one hour and 20 minutes.

Amanda leaves the gym at 1.40pm and cycles home.

She takes 30 minutes.

Her speed on the return journey is 24km/h.

Travelling at a Constant Speed

- When the speed you are travelling at does not change, it remains **constant**.
- You can work out speed, distance or time using a formula triangle.

Example

A car travels 120 miles at 40 miles per hour. How long does the journey take?

time = distance ÷ speed

time = 120 ÷ 40 = 3 hours

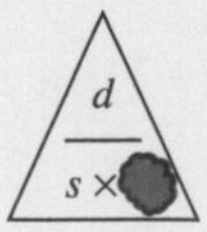

Cover up what you are trying to find.

Key Point

$$\text{speed} = \frac{\text{distance}}{\text{time}}$$

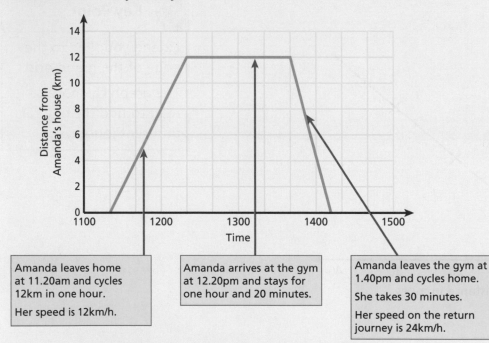

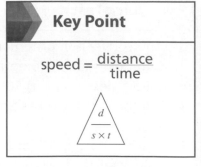

Example
A plane takes $2\frac{1}{2}$ hours to travel 750 miles. What is the speed of the plane?

speed = distance ÷ time

speed = 750 ÷ 2.5

= 300mph

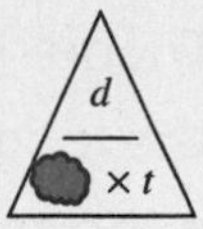

Cover up what you are trying to find.

Unit Pricing

- Unit pricing involves using what 'one' is to work out other amounts. This is often used in conversions of money.

Example
If £1 = \$1.75, how much would a pair of jeans cost in \$ if they were £60?

60 × 1.75 = \$105

How much would a TV costing \$525 be in £?

525 ÷ 1.75 = £300

Multiply to get the dollars; divide to get the pounds.

Density

- You can work out **density**, mass and volume using a formula triangle like speed.

density = mass ÷ volume

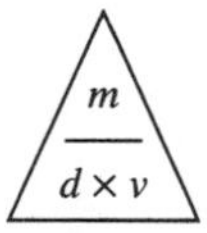

Example
Find the density of an object that has a mass of 60g and a volume of $25cm^3$.

density = mass ÷ volume
= 60 ÷ 25
= $2.4g/cm^3$

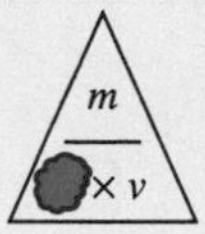

Cover up what you are trying to find.

Quick Test

1. What is 90 minutes in hours?
2. Stuart drives 180km in 2 hours 15 minutes. Work out Stuart's average speed.
3. John travelled 30km in $1\frac{1}{2}$ hours. Kamala travelled 42km in 2 hours. Who had the greater average speed?
4. If £1 = €1.2, what would £200 be worth in €?
5. What is the mass of 250ml of water with density of $1g/cm^3$ ($1000cm^3$ = 1 litre)?

Key Point

Remember to use the correct units.

Volume: $\mathbf{cm^3}$ and $\mathbf{m^3}$

Mass: **g** and **kg**

Density: $\mathbf{g/cm^3}$ and $\mathbf{kg/m^3}$

Key Words

distance
speed
constant
density

Right-Angled Triangles 1

You must be able to:

- Label right-angled triangles correctly
- Know Pythagoras' Theorem
- Find the length of the longest side
- Find the length of a shorter side.

Pythagoras' Theorem

- Remember the formula for **Pythagoras' Theorem:** $a^2 + b^2 = c^2$

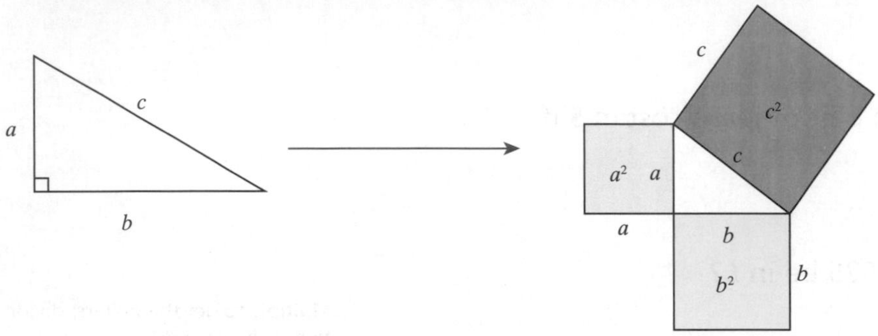

Key Point

The longest side, the hypotenuse, is called c and is opposite the right angle.

The two shorter sides are a and b in any order.

Finding the Longest Side

- To find the longest side (**hypotenuse**), **add** the **squares**. Then take the **square root** of your answer.

Example

Find the length of y. Give your answer to 1 decimal place.

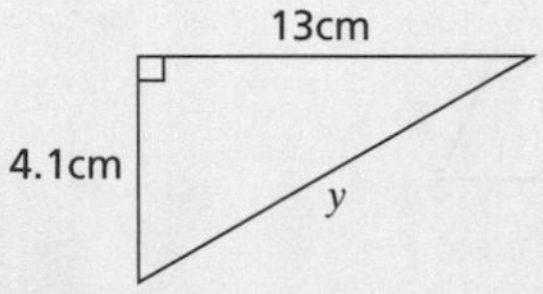

Label the sides a, b and c first.

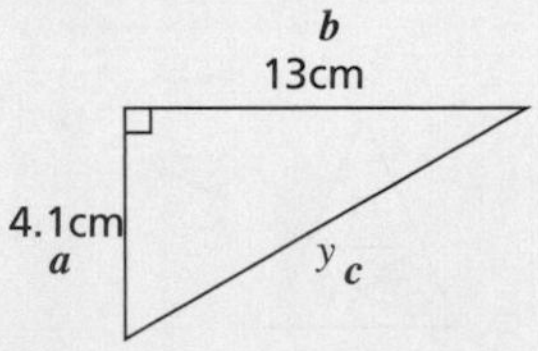

Now use the formula:

$a^2 + b^2 = c^2$

$4.1^2 + 13^2 = y^2$

$16.81 + 169 = y^2$

$185.81 = y^2$

$y = \sqrt{185.81}$

$= 13.6\text{cm (1 d.p.)}$

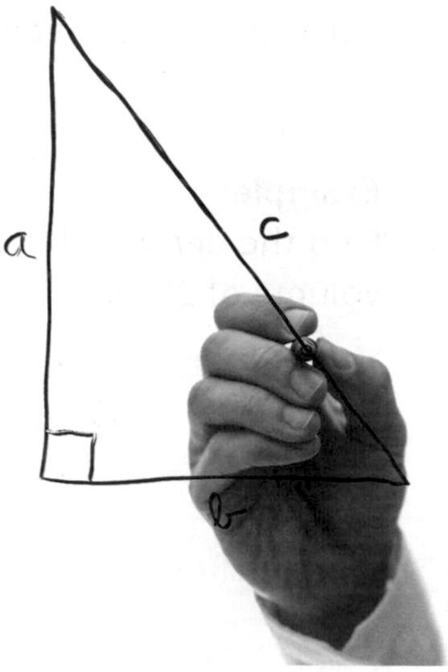

Finding a Shorter Side

- To find a shorter side, **subtract** the squares. Then take the square root of your answer.

Example

Find the length of y. Give your answer to 1 decimal place.

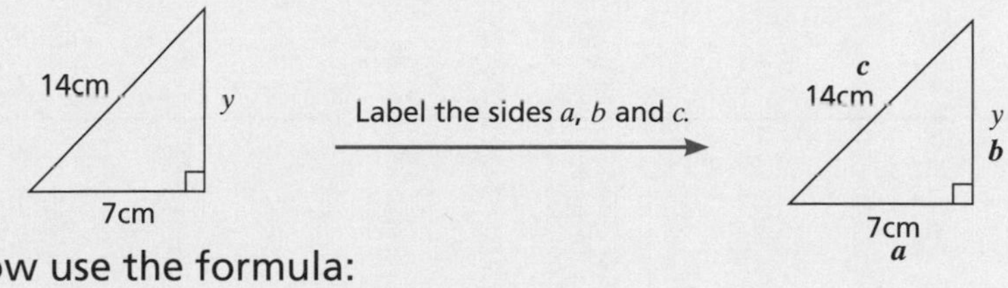

Now use the formula:

$a^2 + b^2 = c^2$

$7^2 + y^2 = 14^2$

$49 + y^2 = 196$

$y^2 = 196 - 49 = 147$

$y = \sqrt{147} = 12.1\text{cm (1 d.p.)}$

Example

Find the length of y. Give your answer to 1 decimal place.

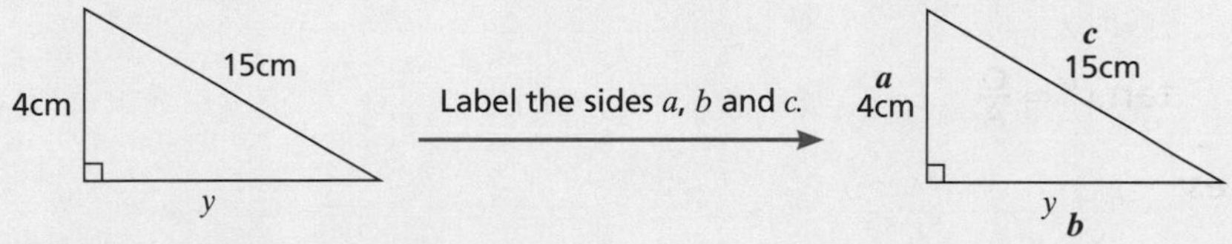

Now use the formula:

$a^2 + b^2 = c^2$

$4^2 + y^2 = 15^2$

$16 + y^2 = 225$

$y^2 = 225 - 16 = 209$

$y = \sqrt{209} = 14.5\text{cm (1 d.p.)}$

Key Point

You will always $\sqrt{}$ at the end.

Quick Test

1. Work out:
   a) 3.2^2
   b) 15.65^2
2. Work out:
   a) $\sqrt{4900}$
   b) $\sqrt{39.69}$
3. Work out the longest side of a right-angled triangle if the shorter sides are 5cm and 2.2cm.
4. Work out the shorter side of a right-angled triangle if the longest side is 12cm and the other shorter side is 9cm.

Key Words

Pythagoras' Theorem
hypotenuse
square
square root

Right-Angled Triangles 2

You must be able to:

- Remember the three ratios
- Work out the size of an angle
- Work out the length of a missing side.

Side Ratios

- Label the sides of the triangle in relation to the angle that is marked.

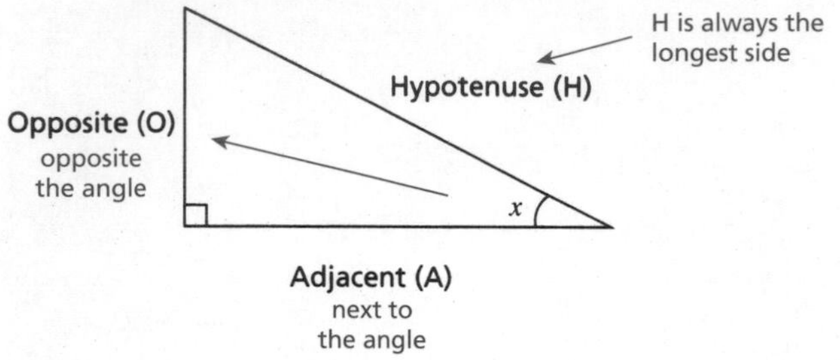

- There are three ratios: sin, cos and tan. Try to find a way of remembering these:

  $\sin x° = \frac{O}{H}$ $\cos x° = \frac{A}{H}$ $\tan x° = \frac{O}{A}$

  – You can use the formula triangles:

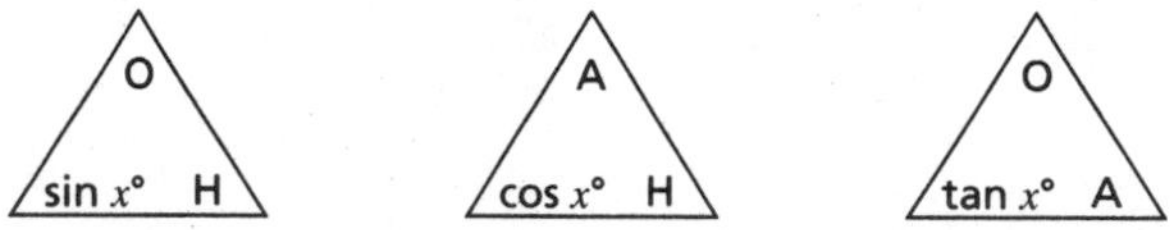

  – You can use a rhyme:
  Some Old Horses Can Always Hear Their Owners Approach

Example
Use your calculator to work out the ratios for these angles:

a) sin 60° = 0.8660
b) cos 45° = 0.7071
c) tan 87° = 19.0811

Example
Use your calculator to work out the angle for these ratios:

a) $\sin x° = 0.5$ $x = \sin^{-1} 0.5 = 30°$
b) $\cos x° = \frac{3}{5}$ $x = \cos^{-1} (3 \div 5) = 53.1°$
c) $\tan x° = 2.9$ $x = \tan^{-1} 2.9 = 71°$

Key Point

Ensure your calculator is in '**degree**' mode.

Finding Angles in Right-Angled Triangles

- You need to know two sides of the triangle to find an angle.

Example

Find the size of angle x. Give your answer to 1 decimal place.

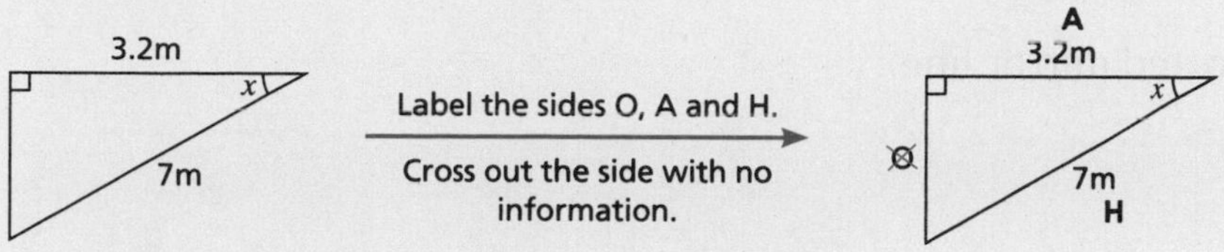
3.2m
x
7m
Label the sides O, A and H.
Cross out the side with no information.
A
3.2m
x
O
7m
H

As **A** and **H** are left we will use the cos ratio.

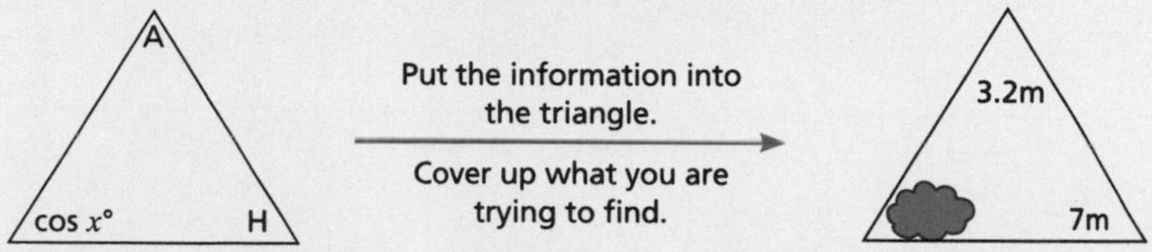
A
cos x°
H
Put the information into the triangle.
Cover up what you are trying to find.
3.2m
7m

$\cos x° = \frac{3.2}{7}$ $\quad x = \cos^{-1}(3.2 \div 7) = 62.8°$

Key Point

The inverse of cos is $\cos^{-1}$.

Finding the Length of a Side

- You need to know the length of one side and an angle to find the length of another side.

Example

Find the length of the side labelled y. Give your answer to 1 decimal place.

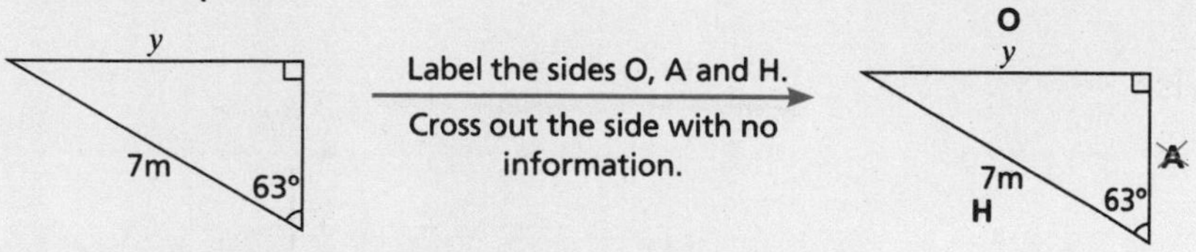
y
7m
63°
Label the sides O, A and H.
Cross out the side with no information.
O
y
A
7m
H
63°

As **O** and **H** are left we will use the sin ratio.

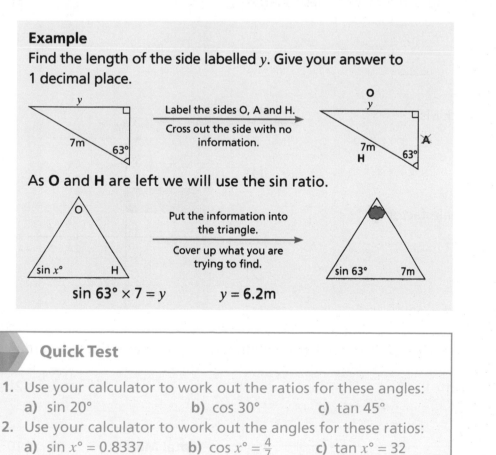

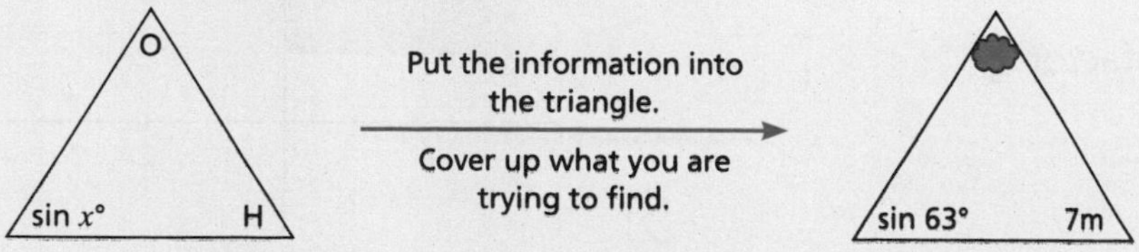
O
sin x°
H
Put the information into the triangle.
Cover up what you are trying to find.
sin 63°
7m

$\sin 63° \times 7 = y$ $\quad y = 6.2\text{m}$

Quick Test

1. Use your calculator to work out the ratios for these angles:
   a) $\sin 20°$ **b)** $\cos 30°$ **c)** $\tan 45°$
2. Use your calculator to work out the angles for these ratios:
   a) $\sin x° = 0.8337$ **b)** $\cos x° = \frac{4}{7}$ **c)** $\tan x° = 32$

Key Words

sin
cos
tan

Review Questions

Symmetry and Enlargement

PS **1** Reflect shape A across the dotted mirror line.

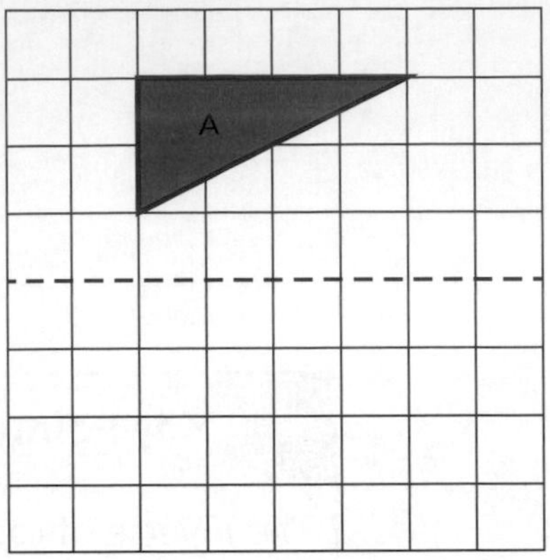

[2]

Total Marks .... / 2

PS **1** What is the order of rotational symmetry of these shapes? [3]

PS **2** **a)** Rotate shape A 90° clockwise about the point (1, 0).

Label the new shape B. [2]

b) Enlarge shape A by a scale factor 2 from the point (–3, 3).

Label the new shape C. [2]

c) Which shapes are congruent? [1]

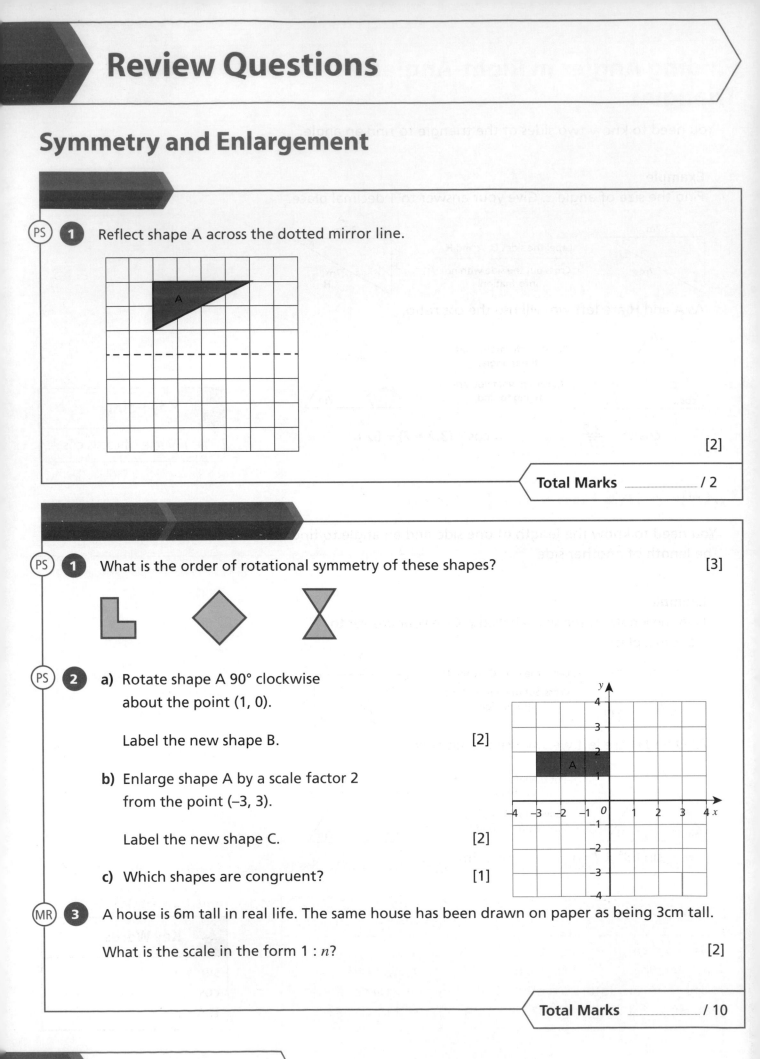

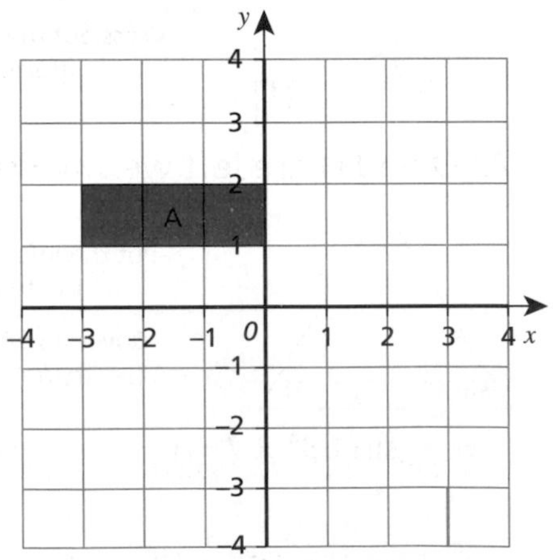

MR **3** A house is 6m tall in real life. The same house has been drawn on paper as being 3cm tall.

What is the scale in the form 1 : n? [2]

Total Marks .... / 10

Ratio and Proportion

(PS) **1** There are 14 boys and 16 girls in a class.

What is the ratio of girls to boys? Write your answer in its lowest terms. [2]

(PS) **2** Simplify the following ratios.

a) 10 : 2 [1]

b) 16 : 24 [1]

c) 25 pence : £2 [2]

Total Marks .... / 6

1 Work out the missing ratio.

a) 8 : 3 = ? : 15 [1]

b) 7 : ? = 63 : 108 [1]

(FS) **2** A sum of money is shared in the ratio 1 : 4

If the larger share is £120, how much money is there altogether? Show your working. [3]

3 Share 40 pens in the ratio 3 : 5, showing your working. [2]

(FS) **4** Danesh bought 18 postcards for £2.16

How much would he pay if he bought 27 postcards? Show your working. [2]

(PS) **5** A telegraph pole 60 feet high casts a shadow 12 feet long. At the same time of day, how long is the shadow cast by a 90-foot pole? [2]

Total Marks .... / 11

Practice Questions

Real-Life Graphs and Rates

1. Concorde could travel 20 miles every minute.

   How many miles per hour (mph) is that? [2]

2. Use £1 = €1.19 to work out how much £3.50 is in euros. [2]

Total Marks .... / 4

1. Each year, there is a tennis competition in Australia and another one in France. The table shows how much money was paid to the winner of the men's competition in each country in one particular year.

Country	Money
Australia	1 000 000 Australian dollars (£1 = 2.70 Australian dollars)
France	780 000 euros (£1 = 1.54 euros)

Which country paid more money? You must show your working. [2]

PS 2. The graph shows the flight details of an aeroplane travelling from London to Madrid, via Brussels.

   a) What is the aeroplane's average speed from London to Brussels? [2]

   b) At what time did the plane arrive in Madrid? [1]

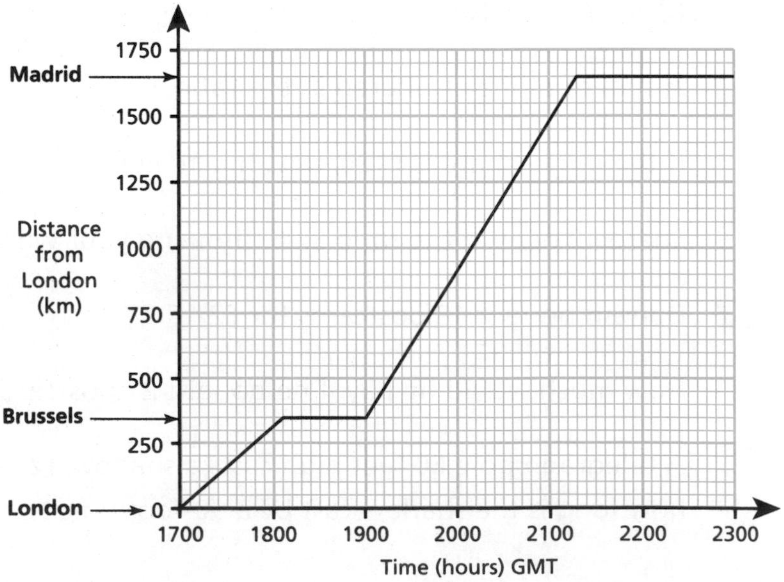

Total Marks .... / 5

Right-Angled Triangles

PS **1** Use your calculator to work out the value of:

a) 17^2 [1]

b) 3.5^2 [1]

c) $\sqrt{529}$ [1]

d) $\sqrt{40.96}$ [1]

Total Marks .... / 4

PS **1** Use Pythagoras' Theorem to work out:

a) the length of AC.

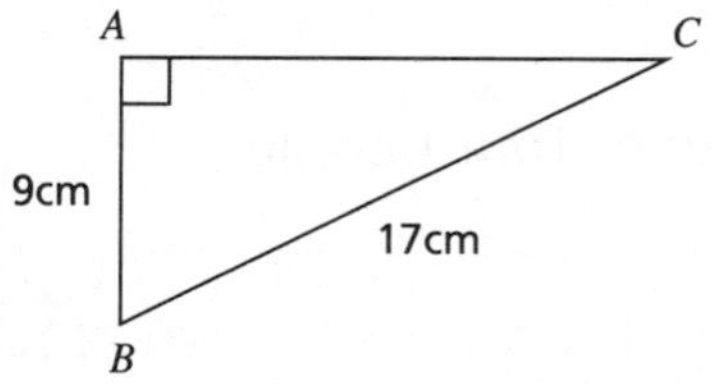

b) the length of BC.

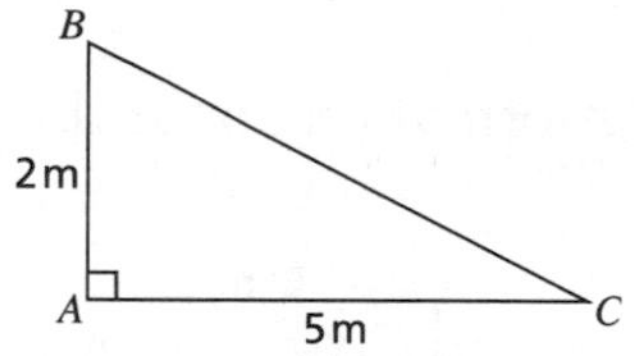

[4]

PS **2** **a)** Work out the value of P.

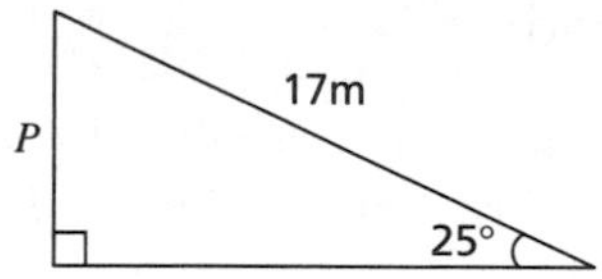

b) Work out the size of angle y.

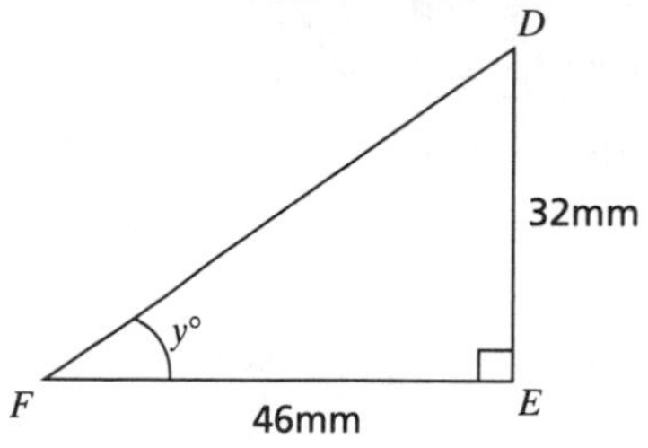

[4]

3 Work out the size of angle x.

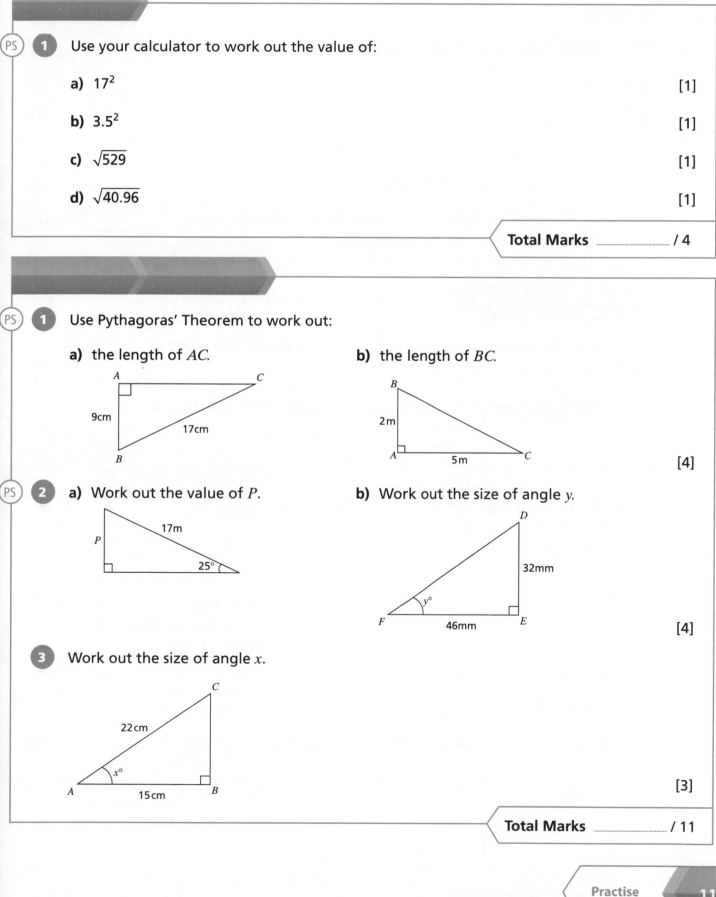

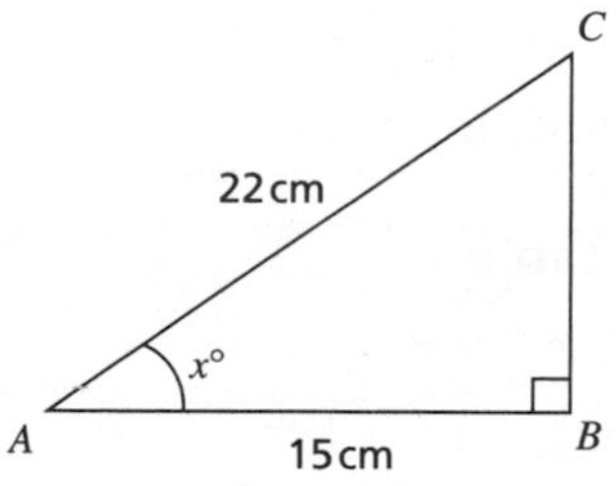

[3]

Total Marks .... / 11

Review Questions

Real-Life Graphs and Rates

(FS) **1** Use £1 = US$1.75 to work out how much:

a) $200 is in £ [2]

b) £200 is in US$ [2]

(PS) **2** A coach travels 300 miles at an average speed of 60mph.

a) For how many hours does the coach travel? [2]

b) At the same speed, how far will the coach travel in four hours? [2]

Total Marks .... / 8

(PS) **1** Calculate the density of a piece of metal that has a mass of 2000kg and a volume of $5m^3$. [2]

2 The graphs show information about the different journeys of four people.

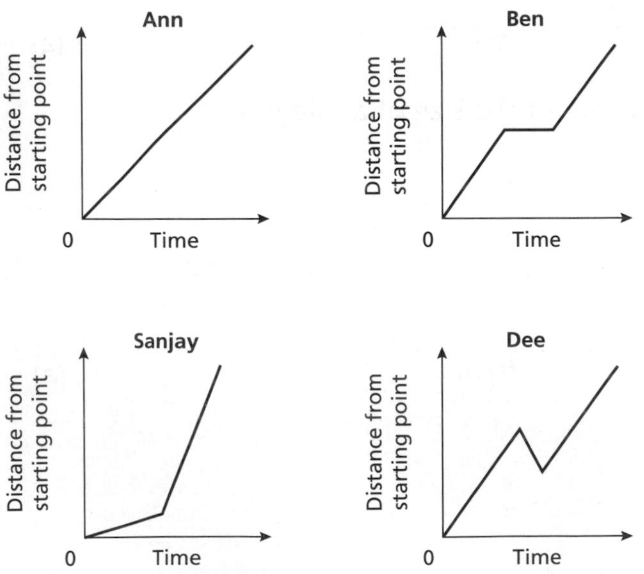

Name	Journey description
	This person walked slowly and then ran at a constant speed.
	This person walked at a constant speed but turned back for a while before continuing.
	This person walked at a constant speed without stopping or turning back.
	This person walked at a constant speed but stopped for a while in the middle.

Write the correct names next to the journey descriptions in the table. [2]

Total Marks .... / 4

Right-Angled Triangles

(PS) **1** Use your calculator to work out the value of:

a) 3.3^2 .... [1]

b) $\sqrt{196}$ .... [1]

(PS) **2** Use Pythagoras' Theorem to work out:

a) the length of y.

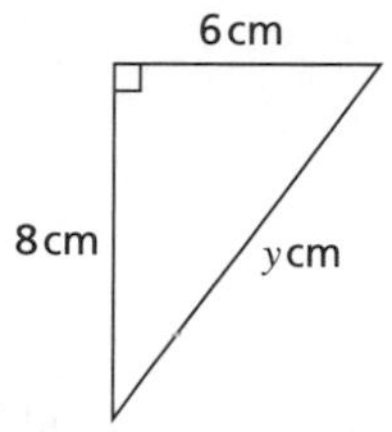

b) the length of x.

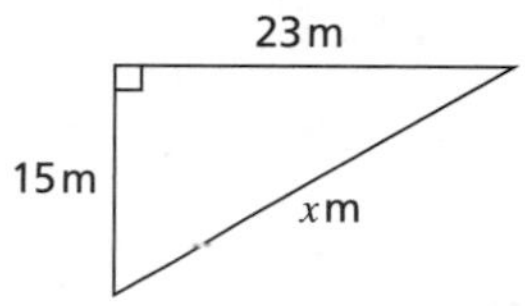

[4]

Total Marks .... / 6

(PS) **1** **a)** Work out the value of angle x.

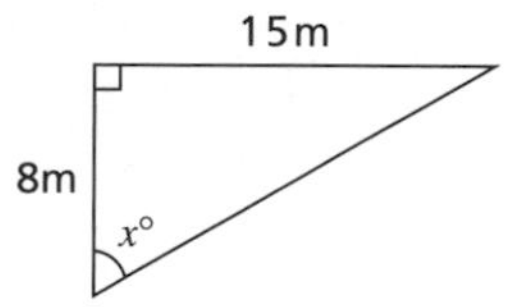

b) Work out the length of AB.

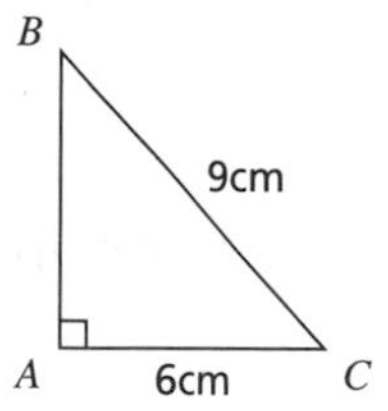

[4]

2 An isosceles triangle has a base of length 4cm and a perpendicular height of 8cm.

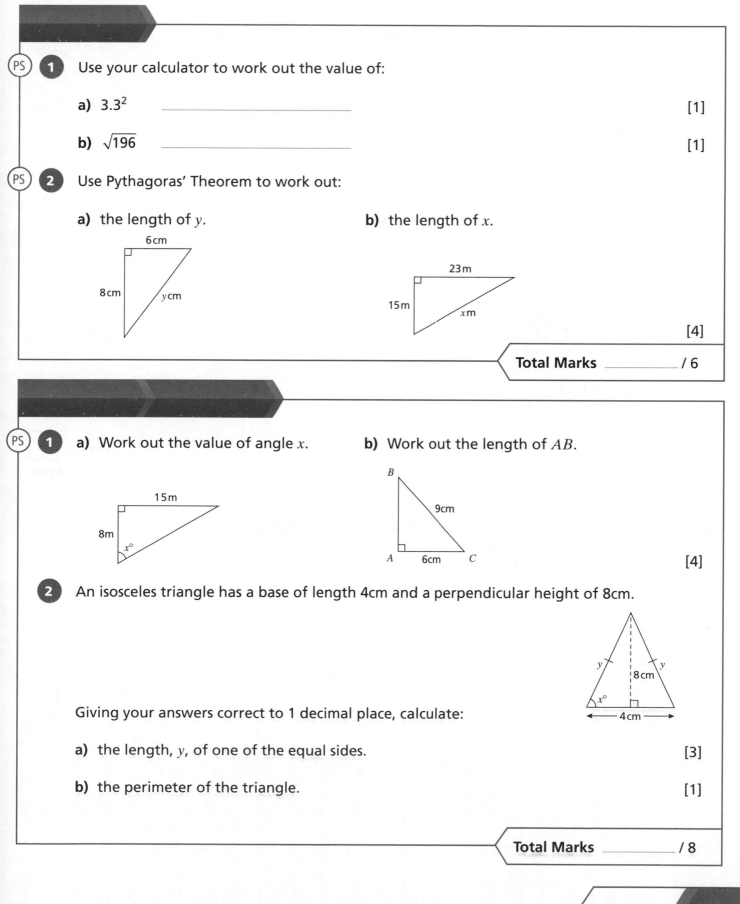

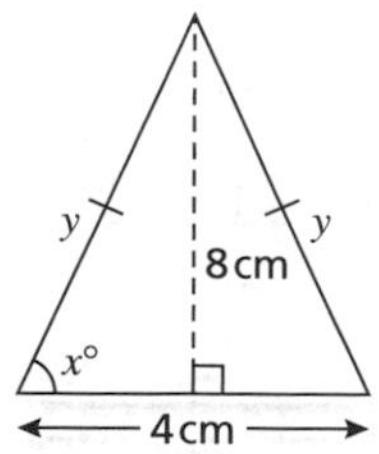

Giving your answers correct to 1 decimal place, calculate:

a) the length, y, of one of the equal sides. [3]

b) the perimeter of the triangle. [1]

Total Marks .... / 8

Mixed Test-Style Questions

No Calculator Allowed

1 Work out both the surface area and volume of each of these cuboids.

a)

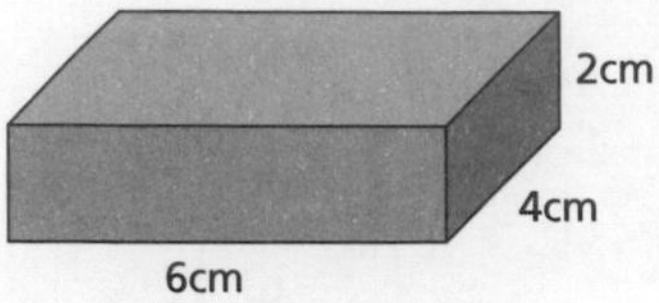

Surface area = ... cm^2

Volume = ... cm^3

b)

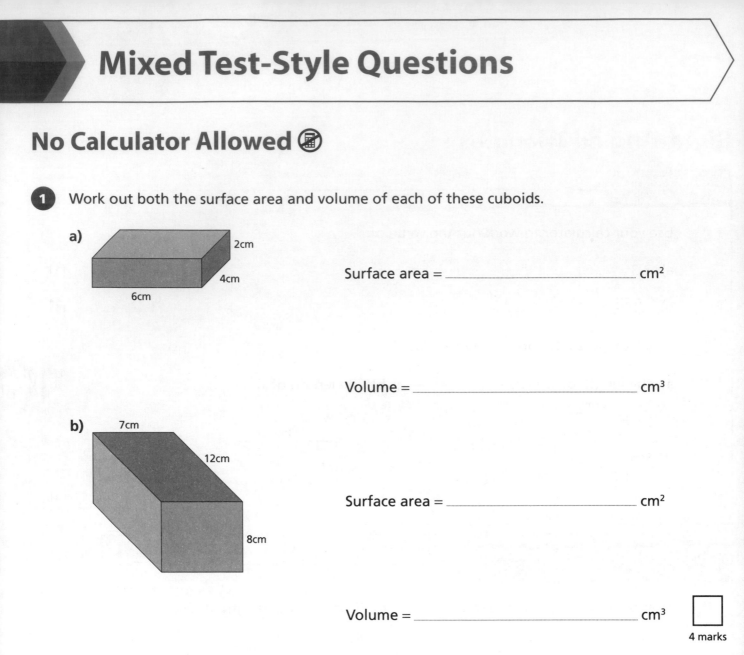

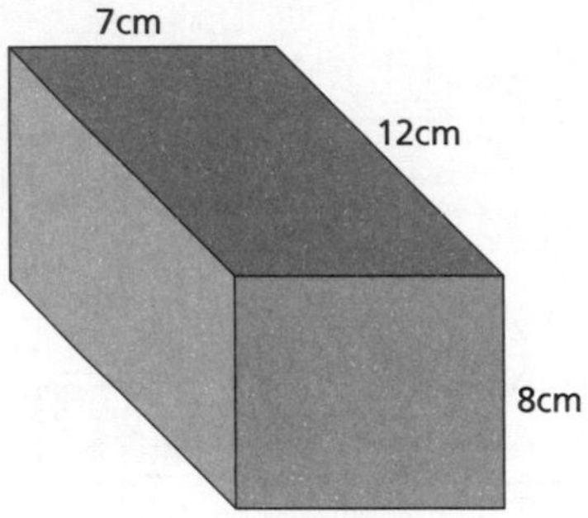

Surface area = ... cm^2

Volume = ... cm^3

4 marks

2 Solve the following, giving your answers in the simplest form.

a) $4\frac{1}{2} + 2\frac{1}{3}$

b) $5\frac{2}{3} + 8\frac{1}{4}$

c) $9\frac{1}{6} - 2\frac{3}{8}$

d) $12\frac{1}{2} - 14\frac{5}{6}$

4 marks

3 On the grid below draw the lines for:

a) $y = 6$ **b)** $x = -4$ **c)** $y = x$

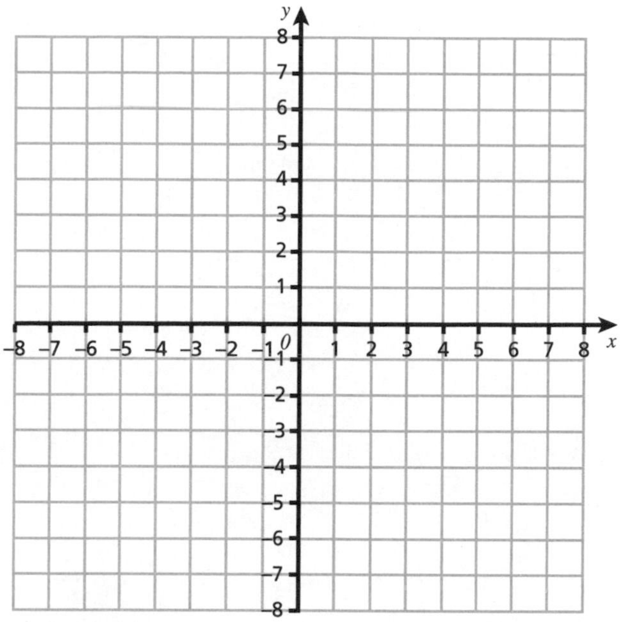

3 marks

4 **a)** Complete the table of coordinates for the equation $y = 3x - 3$.

x	–2	–1	0	1	2	3
y						

b) Plot the coordinates on the graph below and join them with a line.

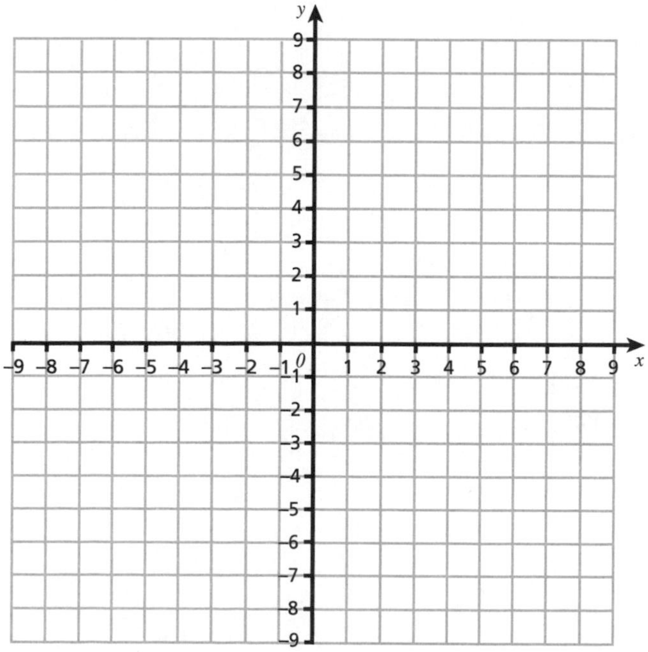

4 marks

TOTAL

15

5 Calculate angles x and y.

a)

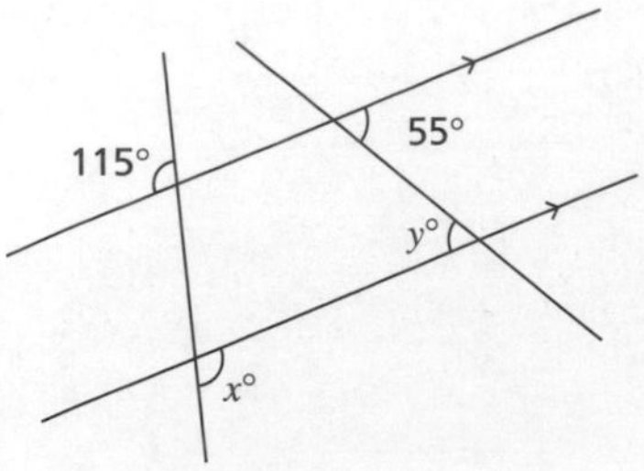

$x =$ .... °

$y =$ .... °

b)

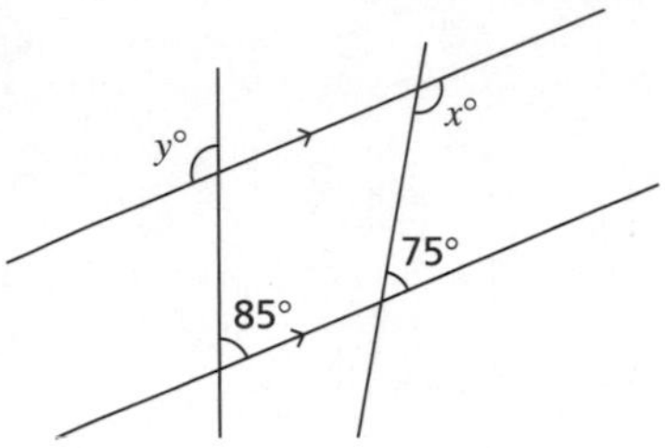

$x =$ .... °

$y =$ .... °

4 marks

6 Simplify the following expressions:

a) $3x - 2y + x + 6y$

b) $4g + 5 - g - 4$

2 marks

7 Expand the following expressions:

a) $4(x - 5)$

b) $4x(x + 4)$

2 marks

8 Factorise completely the following expressions:

a) $6x - 12$

b) $4x^2 - 8x$

2 marks

9 The rectangle and trapezium below have the same area.

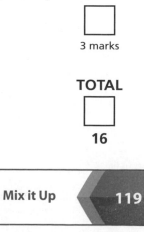

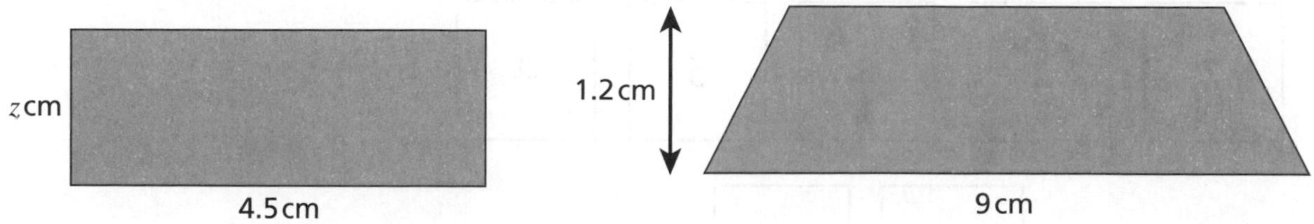

Work out the value of z. Show your working.

3 marks

10 A chocolate bar costs 60p. A vending machine which sells the chocolate bars is emptied and the following coins are found:

Coins	Frequency
£1	25
50p	59
20p	72
10p	31

How many chocolate bars were sold?

3 marks

TOTAL

16

Mixed Test-Style Questions

11 Use the cards below to make two-digit numbers as asked. The first one has been done for you.

7	1	6	5	4

An odd number	7	1
a) A square number		
b) A multiple of 3		
c) A prime number		

2 marks

12 A plant grows by 10% of its height each day. At 8am on Monday the plant was 400mm high.

How tall was it:

a) at 8am on Tuesday?

b) at 8am on Wednesday?

2 marks

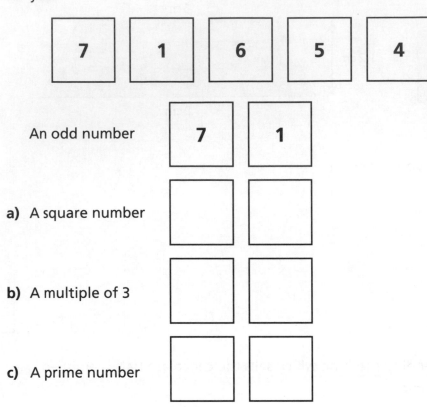

13 What are the names of the following shapes?

a)

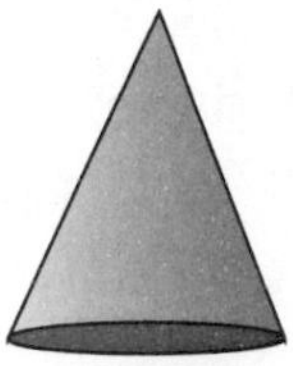

....

b)

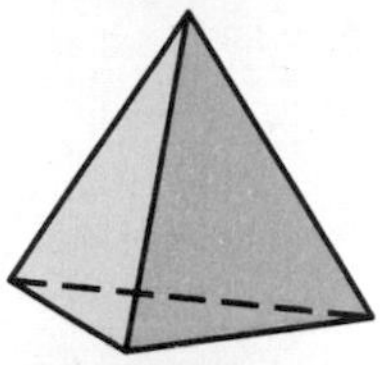

....

c)

....

3 marks

14

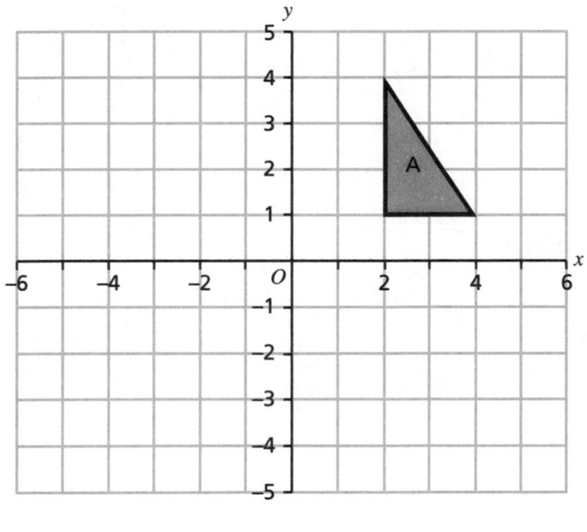

a) Reflect shape A in the y-axis.

b) Enlarge shape A by a scale factor 2 from the point (3, 4).

c) Rotate shape A 180° about (0, 0).

3 marks

TOTAL

10

Mixed Test-Style Questions

Calculator Allowed

1 Sam was sitting on the dock of the bay watching boats for an hour. He collected the following information:

Type	Frequency	Probability
tug boat	12	
ferry boat	2	
sail boat	16	
speed boat	10	

a) Complete the table's probability column, giving your answers as fractions.

b) If Sam saw another 75 boats, estimate how many of them would be sail boats.

5 marks

2 Work out the surface area and volume of these cylinders.

a) radius = 4cm

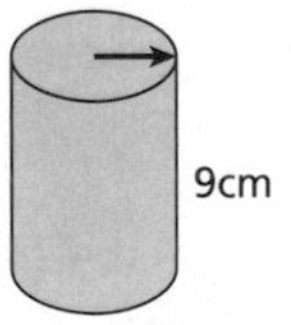

Surface area = ... cm^2

Volume = ... cm^3

b) diameter = 10cm

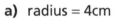

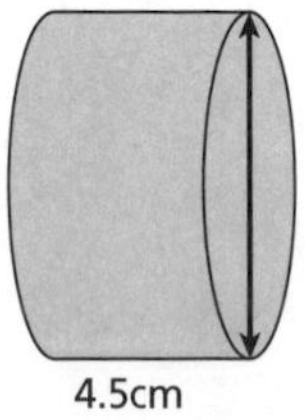

Surface area = ... cm^2

Volume = ... cm^3

8 marks

3 The diagram shows a circle inside a square of side length 4cm.

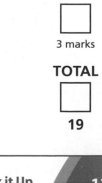

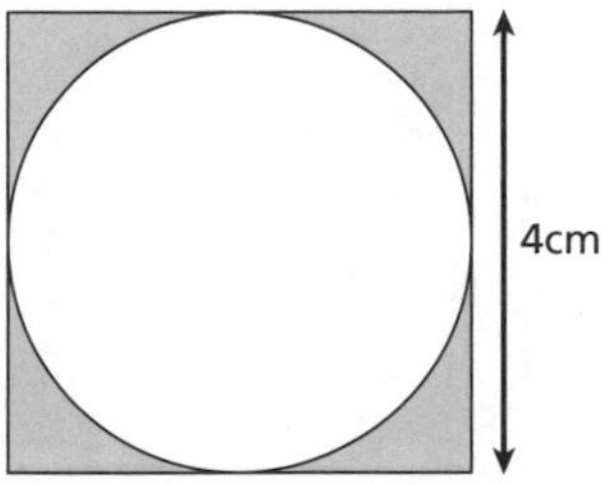

Find the total area of the shaded regions.

.... cm^2

3 marks

4 Barry is planning to buy a car. He visits two garages which have the following payment options:

Mike's motors	**Carol's cars**
£500 deposit	£600 deposit
36 monthly payments of £150	12 monthly payments of £50
£150 administration fee	24 monthly payments of £200

Which garage should Barry buy his car from in order to get the cheapest deal? Show working to justify your answer.

3 marks

TOTAL

19

5 An athlete can run 100m in 12 seconds.

Work out the athlete's speed in:

a) m/s

..m/s

b) km/h

..km/h ☐ 4 marks

6 A wire 15m long runs from the top of a pole to the ground as shown in the diagram. The wire makes an angle of 45° with the ground.

15m

45°

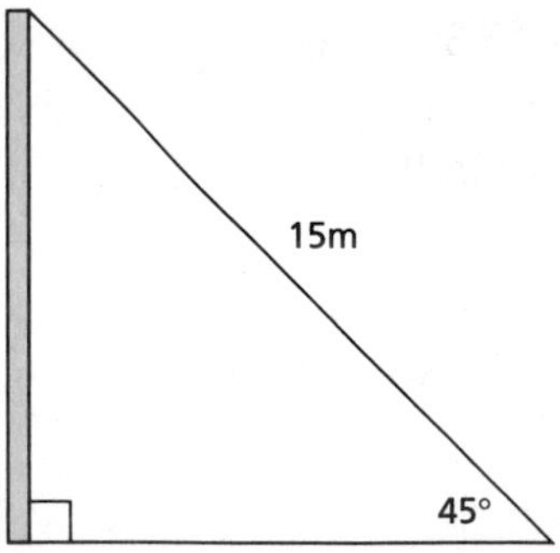

Calculate the height of the pole. Give your answer to a suitable degree of accuracy.

☐ 2 marks

7 Below is a map of an island. The scale is 1cm : 4km.

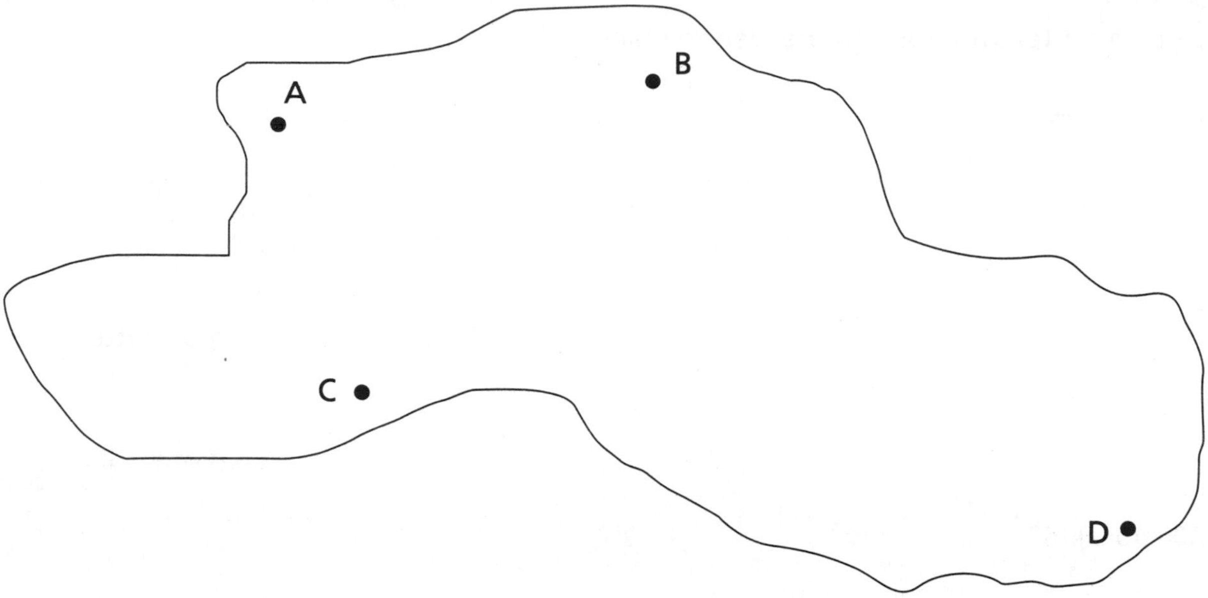

A helicopter flies directly from A to B, B to C, then C to D. What is the total distance flown in km?

..........................km

3 marks

8 Change 8% to:

a) a decimal

b) a fraction in its lowest terms

2 marks

TOTAL

11

9 A recipe for 12 cupcakes needs 80g of butter and 200g of flour.

How much butter and flour are needed to make:

a) 24 cupcakes?

..........g of butter

..........g of flour

b) 30 cupcakes?

..........g of butter

..........g of flour

4 marks

10 Work out the missing numbers. You can use the first line to help you.

$16 \times 21 = 336$

a) $16 \times$ .... $= 112$

b) $336 \div 21 =$ ....

2 marks

11 The data below represents the waiting time in minutes of 15 patients in a doctor's surgery:

49	23	34	10	28	28	25	45
39	35	15	14	48	10	20	

a) Draw a stem-and-leaf diagram to show this information.

b) Use your diagram to find the median waiting time.

3 marks

TOTAL

9

Answers

Pages 6–13 Revise Questions

Page 7 Quick Test

1. 3601, 3654, 3750, 3753, 3813
2. 226 635
3. 163
4. 40
5. $-5 < 3$

Page 9 Quick Test

1. 49
2. 7
3. $2^3 \times 5$
4. 252
5. 8

Page 11 Quick Test

1. Output 4 Input 48
2. 21, 25, 29, 33, 37
3. $\div 2$ **OR** $\times \frac{1}{2}$
4. B
5. 39

Page 13 Quick Test

1. 8, 13, 18, 23, 28
2. 5, 8, 13, 20, 29
3. **a)** $3n + 3$ **OR** $3(n + 1)$ **b)** 153
4. Position to term rule

Pages 14–15 Review Questions

Page 14

1. 1996 **[1]** because 2000 – 1996 = 4 but 2007 – 2000 = 7 **[1]**
2. 245 **[2]**
   OR Method mark for valid attempt to subtract **[1]**

	4	7	6
–	2	3	1
	2	4	5

3. 5000, 46 000, 458 000, 46 000 All 4 correct **[2]** Any 3 correct **[1]**
4. $\frac{27}{50}$, 55%, 0.56, 0.6, 0.63
   All 5 correct **[3]** **OR** 3 out of 5 correct **[2]** **OR** 55% = 0.55 and $\frac{27}{50} = 0.54$ **[1]**
5. Ahmed = 2×22 = 44 years old **[2]** **or** Rebecca = 25 – 3 = 22 years old seen **[1]**
6. 15 878 **[3]**
   OR 12 000 + 1800 + 210 + 1600 + 240 + 28 **[2]**
   OR Valid attempt at multiplication with one numerical error **[1]**
7. 52 **[2]**
   OR Valid attempt at division with one numerical error **[1]**
8. 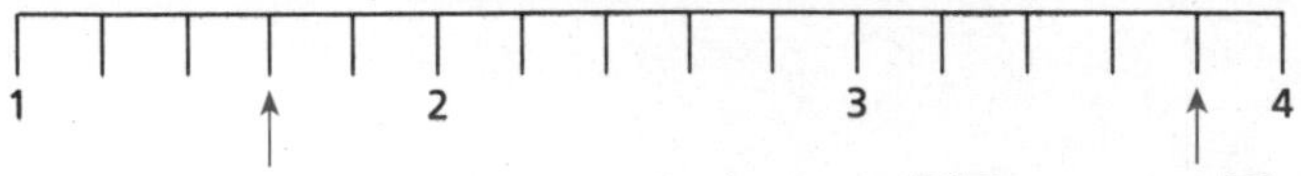

Both correct **[2]** One correct **[1]**

Page 15

1. 200ml **[3]**
   OR 1000ml and 800ml seen **[2]**
   OR 1000ml seen **[1]**
2. **a)** 16 **OR** 36 **[1]**
   b) 13 or 17 or 31 or 37 or 53 or 61 or 67 or 71 or 73 **[1]**
   c) 36 **[1]**
   d) 15 **[1]**
3. 9cm **[2]** **OR** $27 \div 3$ seen **[1]** (an equilateral triangle has three equal sides)
4. 33 **[2]**
   OR 11 seen **[1]**
5. 25° **[3]**
   OR 65 seen **[2]**
   OR 130 seen **[1]**
   (angles in a triangle add up to 180°, base angles in an isosceles triangle are equal, right angle is 90°)
6. $T = 200$ **[1]** ($S = T + 100$, $2T + 100 = 500$) $S = 300$ **[1]**

Pages 16–17 Practice Questions

Page 16

1. **a)** Jan **[1]** **b)** Jul **[1]** **c)** 12 **[1]** (–1 is below zero)
2. $\sqrt{49}$ = 7cm

1. Jessa is right **[1]** Using BIDMAS multiply is first so 3 + 20 + 7 = 30 **[1]**
2. **a)** £4248 **[3]**
   OR 3000 + 500 + 40 + 600 +100 + 8 **[2]**
   OR Valid attempt at multiplication with one numerical error **[1]**
   b) $354 \div 52 = 6.8$ **[2]**
   OR Valid attempt at division with one numerical error **[1]**
   7 coaches needed **[1]**
   c) 10 spare seats **[2]**
   OR 364 seen ($52 \times 7 = 364$) **[1]**
3. $8^2 = 64$ and $9^2 = 81$ so $\sqrt{79}$ is between 8 and 9 **[2]**
   OR attempt to find any two square numbers each side of 79 **[1]**

Page 17

1. **a)** 19 **[1]** 23 **[1]** **b)** + 4 **[1]**
2. **a)** 14 **[1]** **b)** 9 weeks **[1]**

1. **a)** $3n + 1$ **[3]**
   OR $3n$ **[2]**
   OR +3 seen as term to term rule **[1]**
   b) 181 **[1]**
2. 5, 9, 13, 17, 21 = arithmetic. 2, 8, 18, 32, 50 = quadratic.
   8,17, 32, 53, 80 = quadratic. All three correct **[2]** **OR** 1 correct **[1]**

Pages 18–25 Revise Questions

Page 19 Quick Test

1. 24cm
2. 27cm²
3. 6cm²
4. Area 19cm²
   Perimeter 24cm

Page 21 Quick Test

1. 16cm²
2. 13cm²
3. Circumference = 37.7cm (1 d.p.) Area = 113.1cm² (1 d.p.)
4. Circumference = 12.6cm (1 d.p.) Area = 12.6cm² (1 d.p.)

Page 23 Quick Test

1. **a)** Red = 9 Blue = 7 Green = 8 Yellow = 6 Other = 7 **b)** 37
2. **a)** 6 and 3 **b)** mean = 9.2 (1 d.p.) median = 6 range = 37
   c) Median as there is an outlier

Page 25 Quick Test

1. a)

	Football	Rugby	Total
Women	16	9	25
Men	20	10	30
Total	36	19	55

b) 36 c) 9 d) 55

Pages 26–27 Review Questions

Page 26

1. £550 **[3] OR** £2200 ÷ 4 seen **[2] OR** £2200 seen **[1]**
2. Mercury, Venus, Earth, Mars, Jupiter, Saturn, Uranus, Neptune All 8 correct **[2] OR** 6 correct **[1]**
3. 25cm² **[2]**
   OR 5cm seen **[1]**
4. 22 **[2]**
   OR 16 + 6 seen **[1]**

1. 1248 **[1]** (24 is half of 48) 26 **[1]** (26 is half of 52) 48 **[1]**
2. 15 and 12 **[2] OR** either 15 **OR** 12 seen **[1]**
3. 4 packs of sausages and 3 packs of rolls **[3]**
   OR 24 seen **[2]**
   OR Valid attempt to find LCM of 6 and 8 seen **[1]**

Page 27

1. a) 9 **[1]** 16 **[1]**
   b) **[1]** for each correct answer e.g. $n \div 6$ (or equivalent) or $\sqrt{n}$

1. a) 44 **[1]** (for 20th term $n = 20$) **b)** 204 **[1]**
   c) $2n - 1$ **[2] OR** $2n$ seen **[1]**
2. 4.03pm **[3] OR** 180 seen **[2] OR** Valid attempt to split 20 and 45 into prime factors seen **[1]** (This question is asking for the LCM)

Pages 28–29 Practice Questions

Page 28

1. Perimeter = 20cm **[1]** Area = 24cm² **[1]**
2. 32cm² **[2] OR** 8 × 4 seen **[1]**

1. $X = 8$cm **[1]** $Y = 6.8$cm **[1]** (area of a rectangle = $l \times w$)
2. a) 192 **[3] OR** 16 × 12 seen **[2] OR** 120 000 and 625 seen **[1]** (As each tile is 25cm 16 will fit along one side and 12 along the other)
   b) £300 **[2] OR** 20 seen **[1]** (The nearest multiple of 10 bigger than 192 is 200)
   c) 8 tiles **[1]**

Page 29

1. a) (6 + 11 + 9 + 12 + 7) ÷ 5 = 9 **[1]**
   b) 6, 7, 9, 11, 12 numbers in order, 9 is in the middle **[1]**
   c) Any 5 numbers with a mean of 9 **[1]**
2. 20 **[2] OR** 90 seen **[1]**
3. a) An 8 seen as the last number in the 7 row **[1]**
   b) 59 (remember to look at the key) **[1]**
   c) 37 (a stem-and-leaf diagram puts numbers in order) **[1]**

1. a) Phil 66.4 (1 d.p.) **[1]** Dave 68.9 (1 d.p.) **[1]** (add them up and divide by number of values)
   b) Phil 70 **[1]** Dave 175 **[1]**
   c) Dave as higher average **OR** Phil as more consistent **[2]**

Pages 30–37 Revise Questions

Page 31 Quick Test

1. 235.6
2. 5.6781
3. 100 000
4. 16.2, 16.309. 16.34, 16.705, 16.713

Page 33 Quick Test

1. 49.491
2. 17.211
3. 163
4. 150 000

Page 35 Quick Test

1. $7x + 5y + 6$
2. c^2d^2
3. 14

Page 37 Quick Test

1. $8x - 4$
2. $2x - 14y$
3. $5(x - 5)$
4. $2x(x - 2)$

Pages 38–39 Review Questions

Page 38

1. $\pi \times 49$ **[1]** = 153.9cm² **[1]**
2. Area = 45cm² **[1]**, perimeter = 28cm **[1]**
3. Area = $\frac{1}{2}(6 + 8) \times 4$ **[1]** = 28cm² **[1]**

1. (This is a compound area separated into a rectangle and triangle)
   a) 22m² **[3] OR** 20 and 2 seen **[2] OR** only 20 seen **[1]**
   b) 4 tins **[1]**
   c) £48 **[2] OR** answer to **b** × 12 seen **[1]**
2. 5092 **[2] OR** 157.0796 or 1.57 seen **[1]** (this question is about the circumference of a circle; notice the units are different, 50cm = 0.5m)

Page 39

1.

	Boys	Girls
Right-handed	12	14
Left-handed	7	1

Completely correct **[3] OR**
3 boxes correct **[2] OR**
2 boxes correct **[1]**

2. a)

Vegetable	Frequency
Carrot	4
Peas	8
Potatoes	9
Sweetcorn	3

All correct **[2] OR** 3 rows correct **[1]**
   b) Potatoes **[1]** (mode is the most common)

1. a) Median **[1]** as data contains an outlier **[1]** (14 808 much bigger than the rest of the data)
   b) 14 808 **[1]**

Pages 40–41 Practice Questions

Page 40

1. 3.7 × 10 → 37, 3.7 × 100 → 370, 3.7 ÷ 10 → 0.37 and 3.7 ÷ 100 → 0.037. All correct **[2] OR** 1 correct **[1]**
2. 0.5679 greater **[1]** Has 6 hundredths compared to 5 **[1]**

1. a) 57.832 **[1]**
   b) 21.98 **[1]**
   c) 74.154 **[1]**
   d) 216 **[1]**

Page 41

1. $5x = 25$, $x = 9$, $4x = 28$. All correct **[2] OR** 2 correct **[1]**
2. $7y$ **[1]** $12y$ **[1]**

1. Both of them **[1]** $2(x + y)$ expands to $2x + 2y$ **OR** vice versa **[1]**
2. £123 **[3] OR** 48 seen **[2] OR** 120×0.4 seen **[1]**
3. $3x + 7$ **[2] OR** $8x + 2 - 5x + 5$ seen **[1]**
4. $3a(bc + 2)$ **[2] OR** $3(abc + 2a)$ **OR** $a(3bc + 6)$ seen **[1]** (completely means remove all common factors)
5. 144 **[2] OR** 9 seen **[1]** ($ab = a \times b$)
6. $3a - b$, $2a - b$, b. All three correct **[2]** any one correct **[1]**

Pages 42–49 Revise Questions

Page 43 Quick Test

1. Volume = 140cm³
   Surface Area = 166cm²
2. Volume = 440cm³
   Surface Area = 358cm²

Page 45 Quick Test

1. a) Volume = 197.9cm³
   Surface Area = 188.5cm²
   b) Volume = 603.2 cm³
   Surface Area = 402.1cm²
2. 174cm³

Page 47 Quick Test

1. 20°
2. Sunday
3. No Helen still uses more (approx. mean 51).

Page 49 Quick Test

1. e.g. money spent on advertising against sales for that item
2. Own goal
3. Any two from: add a time frame, no overlapping tick boxes, cover all outcomes
4. How many sweets do you usually eat in a week? plus responses boxes that do not overlap e.g. none /1–2 / 3–4 / more than 4

Pages 50–51 Review Questions

Page 50

1. 0.759 **[2] OR** 253×3 seen **[1]**
2. 17 **[2] OR** $85 \div 5$ seen **[1]**
3. 5.2 and 3.8 circled **[1]**
4. 140 **[2] OR** 7000 and 50 seen **[1]**
5. £854.40 **[2] OR** Valid attempt at multiplication with one numerical error **[1]**

1. 6.765, 6.776, 7.675, 7.756, 7.765: all 5 correct **[2] OR** 3 correct **[1]**
2. a) £15.99 **[2] OR** attempt to add up 3 costs with only 1 numerical error **[1]**
   b) £4.01 **[1]**
3. 6.93 **[1]**
4. $-0.5 \leqslant$ error < 0.5 **[1]**

Page 51

1. $11k + 5$ **[1]** $4k + 1$ **[1]**
2. $-2k$ (– sign must be seen) **[1]**
3. cd^2, c^2d, c^2d^2. All three correct **[2] OR** any two correct **[1]**

1. a) ab **[1]** b) $2a + 2b$ or $2(a + b)$ **[1]** c) $3a$ and $5a$ **[1]**
2. $4t(2ut - u + 5)$ **[2]**
3. 3600 metres per hour **[2]** $1200 \div \frac{1}{3}$ **OR** 3 seen **[1]**

Pages 52–53 Practice Questions

Page 52

1.

Cuboid	6	12	8	**[1]**
Square-based pyramid	5	8	5	**[1]**
Hexagonal prism	8	18	12	**[1]**

2. Cylinder **[1]**; triangular prism **[1]**; cube **[1]**
3. a) 150cm³ **[1]**; 190cm² **[1]**
   b) 81cm³ **[1]**; 117cm² **[1]**

1. $942 \div 5^2 \div \pi$ **[1]** = 12.0cm **[1]**
2. $1385 \div 7^2 \div \pi$ **[1]** = 9.00cm **[1]**

Page 53

1. $\frac{360}{45} = 8$

Category	Frequency	Angle	
Brie	21	$21 \times 8 = 168°$	**[1]**
Cheddar	5	$5 \times 8 = 40°$	**[1]**
Stilton	14	$14 \times 8 = 112°$	**[1]**
Other	5	$5 \times 8 = 40°$	**[1]**
Total	45	360°	

Favourite cheese

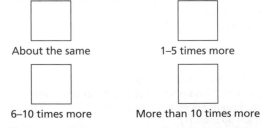

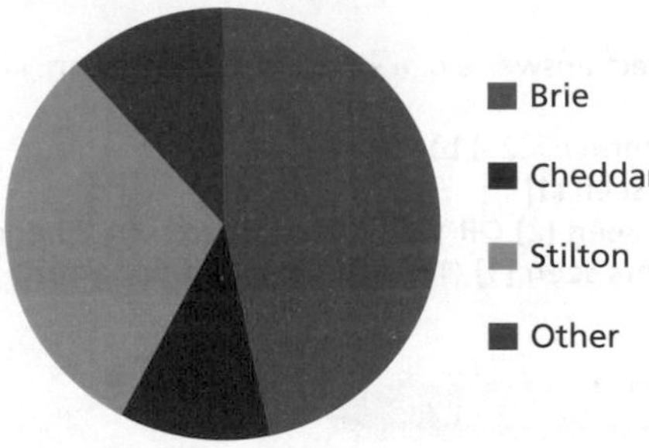

All angles of pie chart correct to within 2° **[1]**; correct labelling **[1]**

2. Monday has around 55 mins. (Add both Andy's and Helen's phone use together). **[1]**

1. Examples could be: Speed of runner against distance, cost of taxi against number of people in taxi, etc. **[2]**
2. How many more times do you go shopping during the Christmas period than other times of the year?

   ☐ About the same ☐ 1–5 times more
   ☐ 6–10 times more ☐ More than 10 times more

   Question should have a time frame (e.g. How many more times do you go shopping during the Christmas period than other times of the year?). **[2]**
   Response boxes should cover all outcomes and not overlap (e.g. about the same; 1–5 times more; 6–10 times more; more than 10 times more). **[2]**
3. a) Scatter graph, frequency polygon, pie chart, bar chart, line grap histogram, etc. Four examples **[2]**
   b) Choose a graph with a reason, for instance pie chart as it shows the percentage of time spent. **[2]**

Pages 54–61 Revise Questions

Page 55 Quick Test

1. e.g. $\frac{4}{6}$ $\frac{6}{9}$ $\frac{8}{12}$ $\frac{10}{15}$ $\frac{12}{18}$
2. $\frac{64}{77}$
3. $\frac{29}{72}$
4. $\frac{15}{52}$
5. $\frac{81}{100}$

Page 57 Quick Test

1. $\frac{1}{3}$
2. $\frac{49}{36} = 1\frac{13}{36}$
3. $\frac{28}{5} = 5\frac{3}{5}$
4. $\frac{37}{4} = 9\frac{1}{4}$
5. $2\frac{17}{90}$

Page 59 Quick Test

1.

x	−2	−1	0	1	2	3
y	−11	−8	−5	−2	1	4

Page 61 Quick Test

1. a) Gradient = 3
   Intercept = 5
   b) Gradient = 6
   Intercept = −7
   c) Gradient = −3
   Intercept = 2
2.

x	−3	−2	−1	0	1	2	3
y	4	2	2	4	8	14	22

Pages 62–63 Review Questions

Page 62

1. Square-based pyramid **[1]**
2. a)

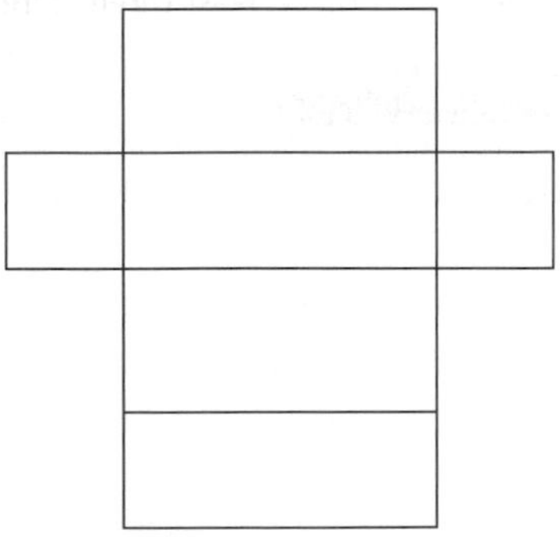

[1]

b)

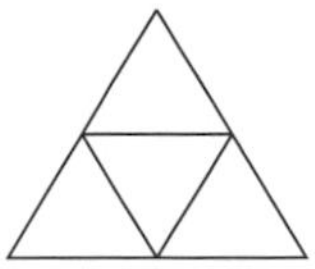

[1]

c)

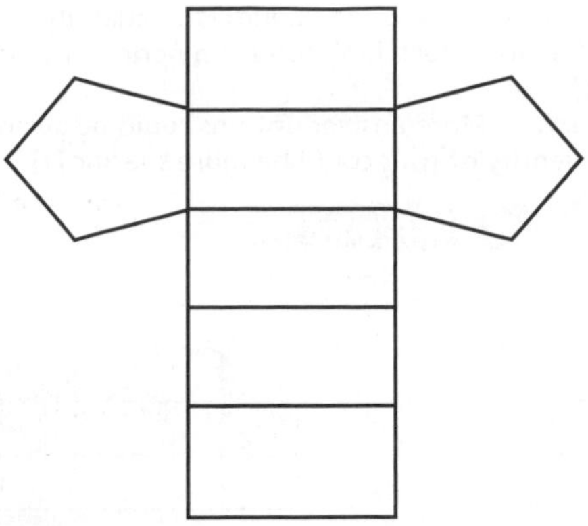

[1]

1. Surface area = 2(14 × 2) + 2(6 × 2) + 2(14 × 6) **[1]** = 248cm² **[1]**
   Volume = 14 × 2 × 6 **[1]** = 168cm³ **[1]**
2. Radius = 2.2 ÷ 2 = 1.1 **[1]**
   $\pi \times 1.1^2 \times 11$ **[1]** = 41.8m³ **[1]**
3. $\sqrt[3]{512}$ **[1]** = 8m = 800cm **[1]**
4. Surface area = 2(15 × 10 + 15 × 20 + 10 × 20) = 1300cm² **[1]**
   No, he has not got enough paper **[1]**

Page 63

1.

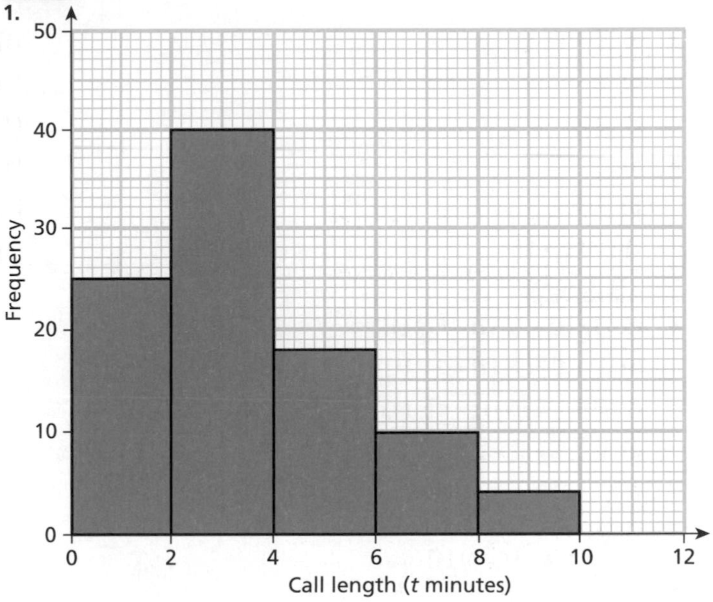

Axes labelled correctly **[1]**; all bars correct **[2]**; three bars correct **[1]**

1.

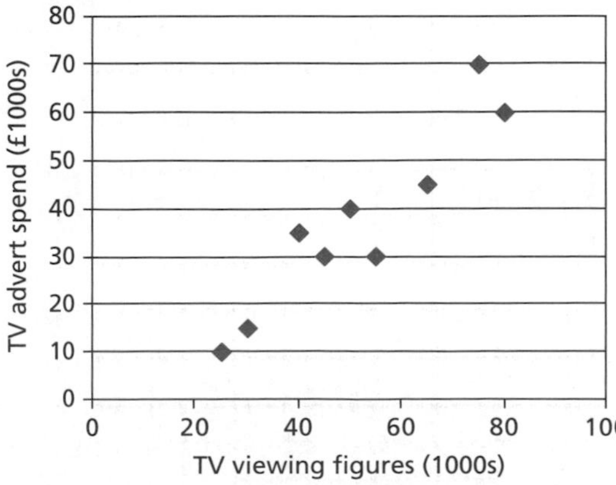

Correct axes labels **[1]**. Points plotted correctly **[1]**. It has a positive correlation **[1]**; the more spent on advertising, the more people are likely to be watching **[1]**

2. a) Example answers: 'a lot' is too vague – the quantity should be specific **[1]**; how much junk food in a period of time could be asked **[1]**
b) Example answers: More answer options could be given, e.g. 0 **[1]**; quantity of fruit could be more specific **[1]**

Pages 64–65 Practice Questions

Page 64

1.

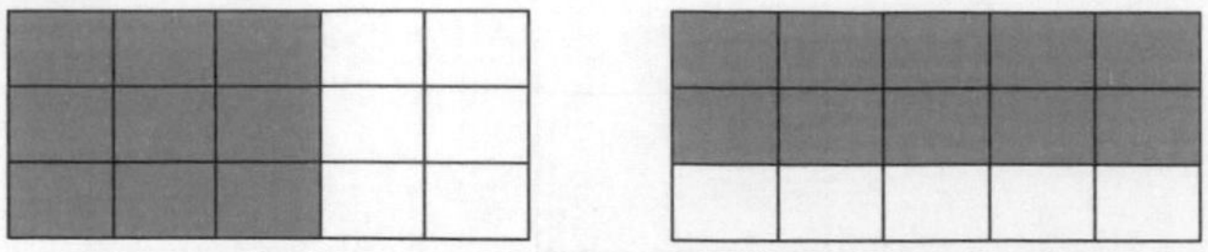

$\frac{3}{5} = \frac{9}{15}$ **[1]** $\frac{2}{3} = \frac{10}{15}$ **[1]**

So $\frac{2}{3}$ is greater than $\frac{3}{5}$ **[1]**

2. a) $\frac{21}{20} = 1\frac{1}{20}$ **[1]**
b) $\frac{41}{40} = 1\frac{1}{40}$ **[1]**
c) $\frac{29}{20} = 1\frac{9}{20}$ **[1]**
d) $\frac{5}{8}$ **[1]**
e) $\frac{3}{10}$ **[1]**
f) $\frac{19}{36}$ **[1]**

3. a) $\frac{2}{24} = \frac{1}{12}$ **[1]**
b) $\frac{40}{54} = \frac{20}{27}$ **[1]**
c) $\frac{3}{20}$ **[1]**

4. a) $\frac{3}{16}$ **[1]**
b) $\frac{9}{48} = \frac{3}{16}$ **[1]**
c) $\frac{21}{12} = \frac{7}{4} = 1\frac{3}{4}$ **[1]**

1. a) $\frac{35}{8} + \frac{11}{5} = \frac{263}{40}$ **[1]** $= 6\frac{23}{40}$ **[1]**
b) $\frac{18}{5} + \frac{21}{9} = \frac{267}{45}$ **[1]** $= 5\frac{14}{15}$ **[1]**
c) $\frac{29}{4} - \frac{30}{11} = \frac{199}{44}$ **[1]** $= 4\frac{23}{44}$ **[1]**
d) $\frac{11}{5} - \frac{10}{7} = \frac{77}{35} - \frac{50}{35}$ **[1]** $= \frac{27}{35}$ **[1]**

Page 65

1. (1, 1) **[1]** and (4, 4) **[1]**
2.

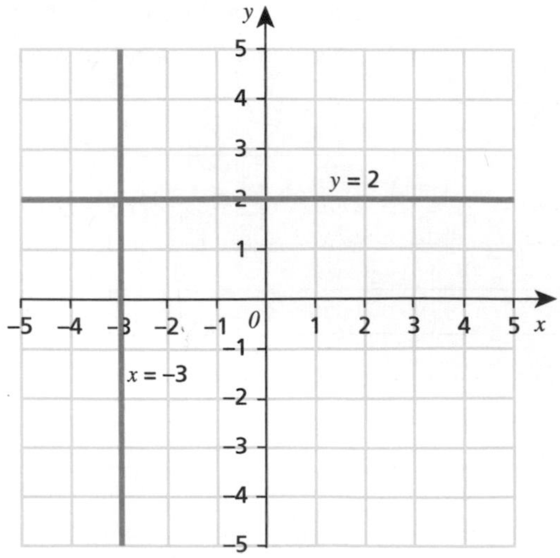

[2]

3.

x	−1	0	1	2	3	4
y	−7	−4	−1	2	5	8

All correct **[2]**; at least four correct **[1]**

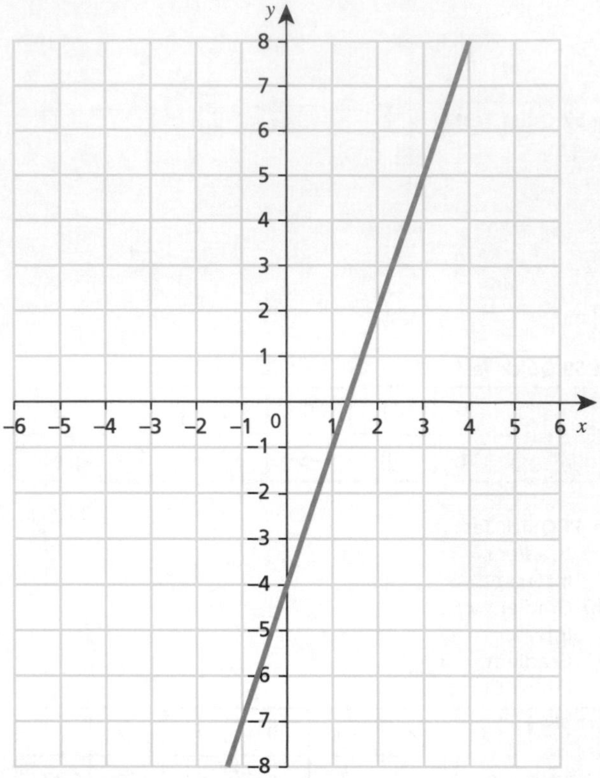

[1]

1.

x	−3	−2	−1	0	1	2	3
y	−2	−4	−4	−2	2	8	16

All correct **[3]**; at least five correct **[2]**; at least three correct **[1]**

Pages 66–73 Revise Questions

Page 67 Quick Test

1. Student's own drawings
2. a) 76°
b) 56°
c) 65°

Page 69 Quick Test

1. a) 55°
b) 112°
c) 126°
2. 1440°
3. Equilateral triangle, square or hexagon

Page 71 Quick Test

1. Likely
2.

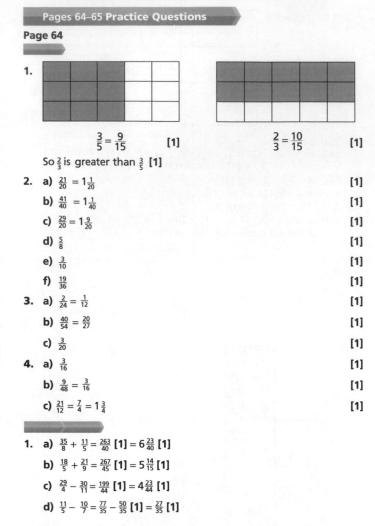

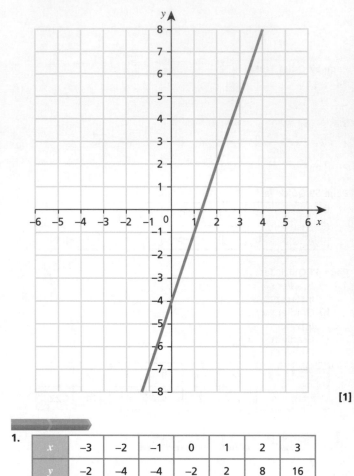

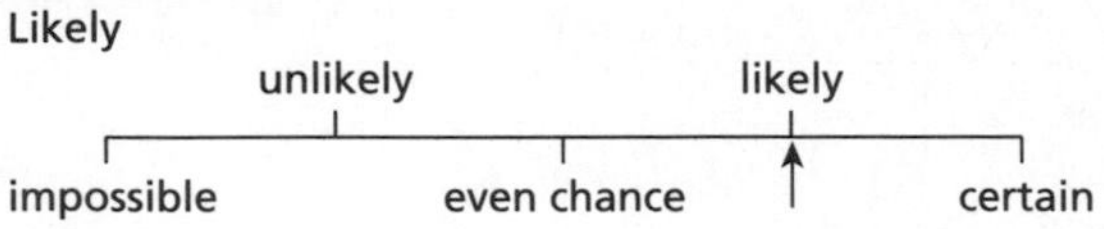

3. a) $\frac{1}{3}$ b) $\frac{2}{3}$
4. 0.15

Page 73 Quick Test

1. a) 0.4
b) 0.6
c) $0.2 \times 60 = 12$

Pages 74–75 Review Questions

Page 74

1. a) $\frac{3}{4} = \frac{6}{8} = \frac{12}{16} = \frac{60}{80}$ etc. **[1]**
   b) $\frac{1}{4} = \frac{2}{8} = \frac{4}{16} = \frac{10}{40}$ etc. **[1]**
   c) $\frac{3}{5} = \frac{6}{10} = \frac{12}{20}$ etc. **[1]**

1. a) $\frac{4}{10} + \frac{1}{10} = \frac{5}{10} = \frac{1}{2}$ **[1]**
   b) $\frac{7}{12} + \frac{3}{12} = \frac{10}{12} = \frac{5}{6}$ **[1]**
   c) $\frac{5}{30} + \frac{6}{30} = \frac{11}{30}$ **[1]**
   d) $\frac{20}{70} + \frac{21}{70} = \frac{41}{70}$ **[1]**
   e) $\frac{8}{9} - \frac{3}{9} = \frac{5}{9}$ **[1]**
   f) $\frac{14}{22} - \frac{11}{22} = \frac{3}{22}$ **[1]**
   g) $\frac{27}{30} - \frac{20}{30} = \frac{7}{30}$ **[1]**
2. a) $\frac{4}{45}$ **[1]**
   b) $\frac{9}{70}$ **[1]**
   c) $\frac{10}{36} = \frac{5}{18}$ **[1]**
   d) $\frac{2}{9} \times \frac{4}{1} = \frac{8}{9}$ **[1]**
   e) $\frac{4}{5} \times \frac{11}{6} = \frac{44}{30} = \frac{22}{15} = 1\frac{7}{15}$ **[1]**
3. $\frac{2}{5} + \frac{1}{4} = \frac{8}{20} + \frac{5}{20} = \frac{13}{20}$ **[1]**
   $1 - \frac{13}{20}$ **[1]** $= \frac{7}{20}$ **[1]**
4. $\frac{4}{9} + \frac{1}{3} = \frac{7}{9}$ **[1]** $1 - \frac{7}{9} = \frac{2}{9}$ **[1]**
   Shared equally $= \frac{1}{9}$ chocolate **[1]**
5. a) $\frac{77}{9}$ **[1]**
   b) $\frac{23}{7}$ **[1]**
   c) $\frac{14}{11}$ **[1]**

Page 75

1. and **2.**

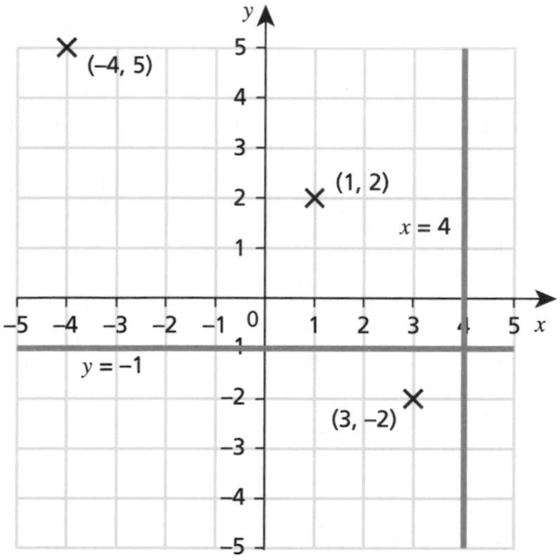

Each point correctly plotted **[1]**; each line correctly drawn **[1]**

1. ×2 → –4

x	–1	0	1	2	3
y	–6	–4	–2	0	2

[2]

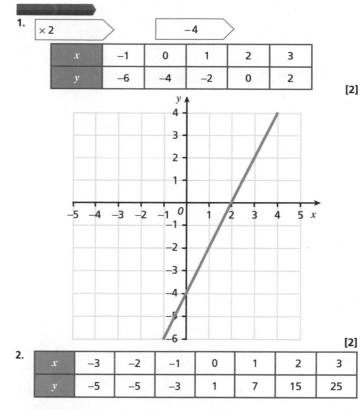

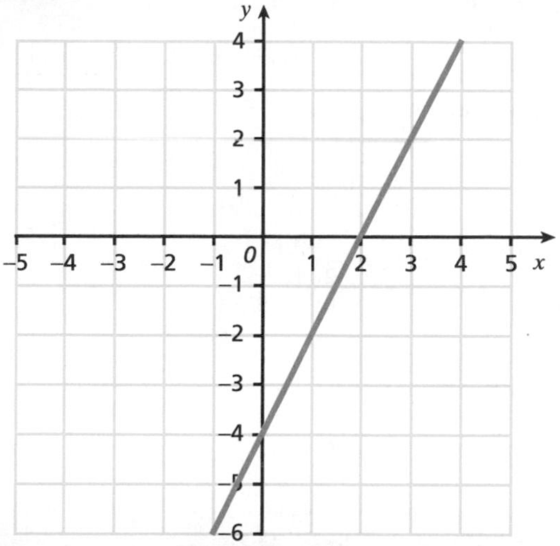

[2]

2.

x	–3	–2	–1	0	1	2	3
y	–5	–5	–3	1	7	15	25

All correct **[3]**; At least five correct **[2]**; At least three correct **[1]**

Pages 76–77 Practice Questions

Page 76

1. a) 180 – 146 = 34° **[1]**
   b) 76° (as angles are opposite) **[1]**
   c) 90 – 56 = 34° **[1]**
2. 180 – 24 = 156° **[1]**, $\frac{156}{2}$ = 78° **[1]**
3. Heptagon **[1]**

1. a) x = 134° **[1]** as alternate (Z angle), y = 180 – 134 = 46° **[1]**
   b) x = 180 – 53 = 127° **[1]**; y = 127° **[1]** as it is a corresponding angle (F angle)
2. 180 – 150 = 30° **[1]** (exterior angle)
   $\frac{360}{30}$ = 12 sides **[1]**

Page 77

1. Impossible or 0 **[1]**
2. a) $\frac{3}{8}$ (number of zeros ÷ total options) **[1]**
   b) $1 - \frac{1}{8}$ **[1]** $= \frac{7}{8}$ **[1]**

1. 1 – 0.65 **[1]** = 0.35 **[1]** (Probability of all outcomes take away probability of it landing other way up.)
2. a)

Number	Frequency	Estimated probability
1	5	$\frac{5}{50} = \frac{1}{10}$
2	8	$\frac{8}{50} = \frac{4}{25}$
3	7	$\frac{7}{50}$
4	7	$\frac{7}{50}$
5	8	$\frac{8}{50} = \frac{4}{25}$
6	15	$\frac{15}{50} = \frac{3}{10}$
	Total 50	1

All correct **[3]**; at least four rows correct **[2]**; at least two rows correct **[1]**

b) i) $\frac{3}{10}$ [1]

ii) $\frac{5}{50} + \frac{7}{50} + \frac{8}{50} = \frac{20}{50} = \frac{2}{5}$ [1]

iii) $\frac{15}{50} + \frac{8}{50} = \frac{23}{50}$ [1]

Pages 78–85 Revise Questions

Page 79 Quick Test

1. 0.35 and 35%
2. $\frac{9}{25}$
3. $28
4. $49

Page 81 Quick Test

1. £405
2. £92 000
3. 84%

Page 83 Quick Test

1. ÷ 6
2. 12
3. 5
4. $y = 5$
5. $x = -9$

Page 85 Quick Test

1. $x = 5$
2. $x = 3$
3. $x = 2$
4. 11
5. 2kg

Pages 86-87 Review Questions

Page 86

1. a) 180 – 90 – 46 = 44° [1]
   b) 180 – 128 = 52° [1]
   c) Because it is an isosceles triangle, 180 – 38 = 142, $\frac{142}{2}$ = 71° [1]

1. a) y = 119° **[1]** (as it is corresponding)
   x = 180 – 61 = 119° **[1]**
   b) y = 180 – 114 = 66° **[1]**
   x = 66° **[1]** (as it is an alternate angle)
2. 1260° [1]
3. 180 – 160 = 20° **[1]**
   $\frac{360}{20}$ = 18, so it is an 18-sided shape. **[1]**

Page 87

1. From left: impossible, unlikely, even chance, likely, certain
   All correct **[2]**; three correct **[1]**
2. a) $\frac{3}{10}$ [1]
   b) $\frac{4}{10} + \frac{2}{10} = \frac{6}{10} = \frac{3}{5}$ [1]
   c) $1 - \frac{1}{10} = \frac{10}{10} - \frac{1}{10}$ **[1]** $= \frac{9}{10}$ **[1]**

1. a)

Sprinkles	Frequency	Probability
Chocolate	19	$\frac{19}{50} = 0.38$
Hundreds and thousands	14	$\frac{14}{50} = 0.28$
Strawberry	7	$\frac{7}{50} = 0.14$
Nuts	10	$\frac{10}{50} = 0.2$

All correct **[2]**; at least two rows correct **[1]**

b) $\frac{19}{50} + \frac{10}{50}$ **[1]** $= \frac{29}{50}$ or 0.58 **[1]**

2. 1 – 0.47 = 0.53 [1]
3. a)

Sales destination	Probability of going to destination
London	0.26
Cardiff	0.15
Chester	0.2
Manchester	0.39

[1]

b) Cardiff (it has the lowest probability) [1]

Pages 88–89 Practice Questions

Page 88

1.

Fraction	Decimal	Percentage
$\frac{7}{10}$	0.7	70
$\frac{55}{100} = \frac{11}{20}$	0.55	55
$\frac{32}{100} = \frac{8}{25}$	0.32	32
$\frac{3}{100}$	0.03	3

[1] [1] [1] [1]

2. a) 50% of £32 = 32 ÷ 2 = £16 [1]
   b) 10% of 80cm = 80 ÷ 10 = 8cm [1]
   c) 10% of 160m = 160 ÷ 10 = 16m
   5% = 16 ÷ 2 = 8m [1]
   15% = 16 + 8 = 24m [1]
   d) 50% = 52
   25% = £26 [1]
3. a) £17 [1] b) £80 [1] c) 15m [1]

1. a) 15 ÷ 3 [1] b) 210 ÷ 7 [1] c) 6000 ÷ 5 [1]
   5 × 2 [1] 30 × 3 [1] 1200 × 4 [1]
   = £10 [1] = £90 [1] = £4800 [1]
2. 10% of £75 = 75 ÷ 10 = £7.50
   20% = £7.50 × 2 = £15 [1]
   Sale price = £75 – £15 [1]
   = £60 [1]
3. Karim gets $\frac{16}{20} = \frac{80}{100}$ [1]
   ×5
   = 80% [1]
   John gets $\frac{15}{20} = \frac{75}{100}$ [1]
   ×5
   = 75% [1]

Page 89

1. a) $\triangle$ = 10 [1]
   b) $\triangle$ = 5 [1]
   c) $n = 12$ [1]
   d) $y = 7$ [1]
2. a) $3n + 1 = 13$
   (–1) $3n = 12$ [1]
   (÷3) $n = 4$ [1]
   b) $2x - 5 = 3$
   (+5) $2x = 8$ [1]
   (÷2) $x = 4$ [1]
   c) $5y + 1 = 11$
   (–1) $5y = 10$ [1]
   (÷5) $y = 2$ [1]

1. a) $3x + 1 = x + 7$
   (–x) $2x + 1 = 7$
   (–1) $2x = 6$ [1]
   (÷2) $x = 3$ [1]
   b) $2(2x - 3) = x - 3$
   $4x - 6 = x - 3$
   (–x) $3x - 6 = -3$ [1]
   (+6) $3x = 3$
   (÷3) $x = 1$ [1]
   c) $6(x + 1) = 2(x + 13)$
   $6x + 6 = 2x + 26$
   (–$2x$) $4x + 6 = 26$ [1]
   (–6) $4x = 20$
   (÷4) $x = 5$ [1]

d) $\frac{3x+5}{4} = 5$

(×4) $3x + 5 = 20$ [1]

(–5) $3x = 15$

(÷3) $x = 5$ [1]

2. $3n + 2 = 11$ [1]

(–2) $3n = 9$

(÷3) $n = 3$ [1]

3. $n + 16 = 28$ [1]

(–16) $n = 12$ [1]

Pages 90–97 Revise Questions

Page 91 Quick Test

1. 6
2. a) 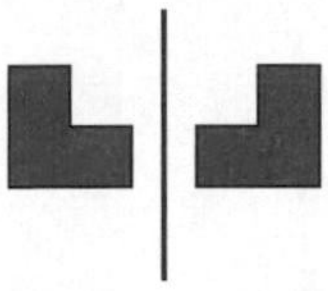 b) 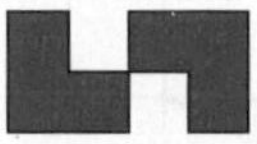 c) 
3. Rectangle 3cm × 6cm

Page 93 Quick Test

1. Any three shapes exactly the same size.
2. D
3. 4–5m
4. A and B
5. Any two similar shapes

Page 95 Quick Test

1. a) 6 : 8 b) 8 : 6
2. a) 1 : 3 b) 7 : 1 c) 1 : 4

Page 97 Quick Test

1. 10 : 25 : 5
2. Sara £200, John £160
3. a) 20 b) 28
4. £40

Pages 98–99 Review Questions

Page 98

1. a) $\frac{3}{20} = \frac{15}{100} = 15\%$ [1]

b) $0.8 = \frac{8}{10}$ [1] $= \frac{4}{5}$ [1]

2. a) 23 ÷ 2 = £11.50 [1]

b) 45 ÷ 10 = 4.5cm [1]

c) 180 ÷ 4 = 45m [1]

d) 10% = 70 ÷ 10 = 7 [1]

5% = 7 ÷ 2 = £3.50 [1]

3. a) £14 [1]

b) £240 [1]

c) £680 [1]

1. a) 5 ÷ 5 × 2 [1]

= £2 [1]

b) £5 – (£2.50 + £2) [1]

= £0.50 or 50 pence [1]

2. 10% of £90 = 90 ÷ 10 = £9

5% = £9 ÷ 2 = £4.50 [1]

15% = £9 + £4.50 = £13.50 [1]

Sale price = £90 – £13.50

= £76.50 [1]

3. 150 ÷ 100 × 6

6% = £9 [1]

After four years £9 × 4 = £36 [1]

Total in account = £186 [1]

Page 99

1. a) ◯ = 4 [1]

b) $n = 9$ [1]

c) $p = 18$ [1]

d) ◯ = 19 [1]

2. a) $4n - 1 = 11$

(+1) $4n = 12$ [1]

(÷4) $n = 3$ [1]

b) $5x + 1 = 21$

(–1) $5x = 20$ [1]

(÷5) $x = 4$ [1]

c) $3a + 8 = 5$

(–8) $3a = -3$ [1]

(÷3) $a = -1$ [1]

1. a) $6x - 5 = 4x + 7$

($-4x$) $2x - 5 = 7$

(+5) $2x = 12$ [1]

(÷2) $x = 6$ [1]

b) $5(x + 2) = 2(x - 1)$

$5x + 10 = 2x - 2$

($-2x$) $3x + 10 = -2$ [1]

(–10) $3x = -12$

(÷3) $x = -4$ [1]

c) $3x - 1 = 4 - 2x$

($+2x$) $5x - 1 = 4$ [1]

(+1) $5x = 5$

(÷5) $x = 1$ [1]

2. $\frac{20n + 150}{5} = 50$ [1]

(×5) $20n + 150 = 250$

(–150) $20n = 100$

(÷20) $n = 5$

The builders worked for 5 hours. [1]

3. $56 - n = 29$ [1]

$n = 56 - 29 = 27$

27 chocolate bars were sold. [1]

Pages 100–101 Practice Questions

Page 100

1. 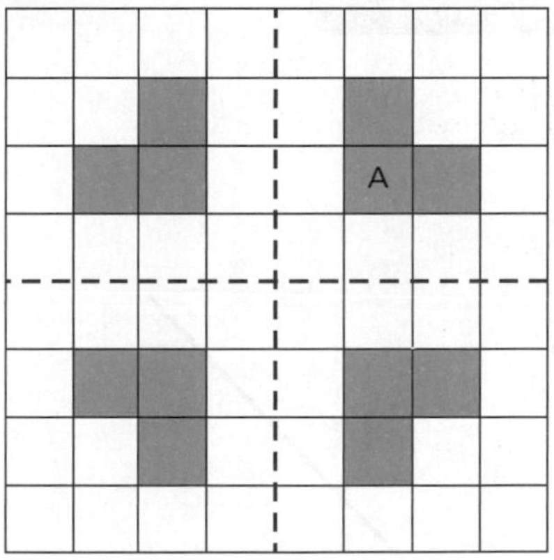

[3]

2. Order 2 [1]

1. 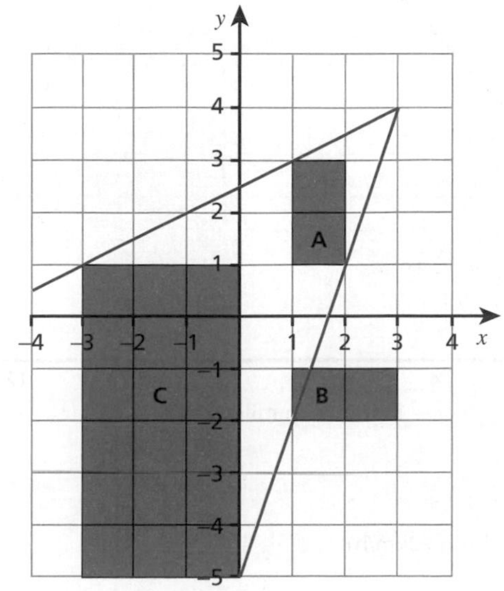

a) See diagram [2]
b) See diagram. [2]
c) A and B [1]

2. 5cm × 4 = 20cm [1]
7cm × 4 = 28cm [1]

Page 101

1. 3 : 6 **OR** 1 : 2 [1]
2. a) 1 : 3 [1]
b) 6 : 1 [1]
c) 20cm : 100cm [1]
1 : 5 [1]
d) 80 mins : 90 mins [1]
8 : 9 [1]

1. 450 ÷ 9 = 50 [1]
Ann: 4 × 50 = £200 [1]
Ben: 5 × 50 = £250 [1]
2. 3 parts = £27
1 part = 27 ÷ 3 = £9 [1]
2 parts = £9 × 2 = £18 [1]
Total sum of money = £27 + £18 = £45 [1]
3. Butter: 40 ÷ 6 × 12 = 80g [1]
Flour: 100 ÷ 6 × 12 = 200g [1]
4. a) 2.5 × 50 000 = 125 000cm [1]
= 1250m
= 1.25km [1]
b) 1.4 × 50 000 = 70 000cm [1]
= 700m
= 0.7km [1]

Pages 102–109 Revise Questions

Page 103 Quick Test

1. a) 40km b) 16km
2. a) 19 miles b) 25 miles
3. a) $3 b) $4
4.

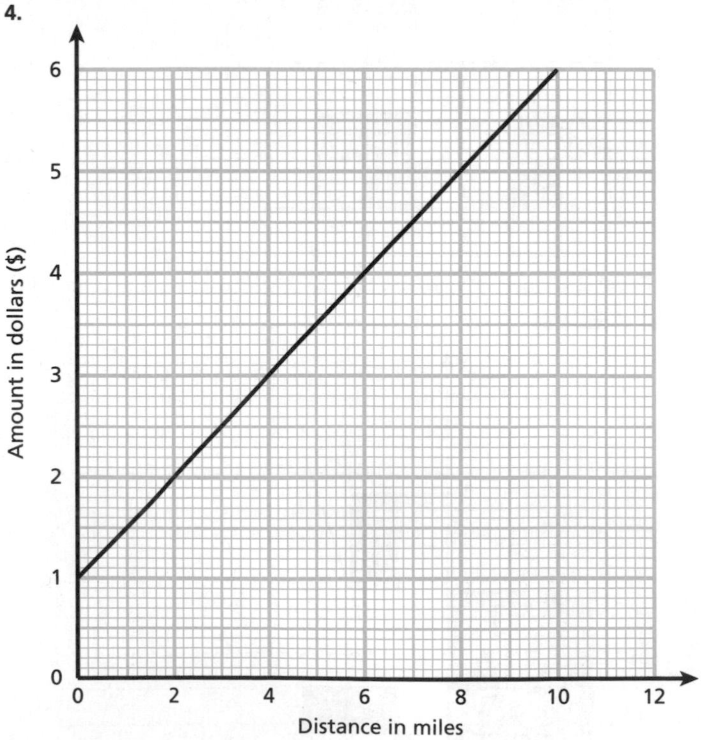

Page 105 Quick Test

1. 1.5 hours
2. 80km/h
3. Kamala (21km/h, John 20km/h)
4. €240
5. 250g

Page 107 Quick Test

1. a) 10.24 b) 244.9225
2. a) 70 b) 6.3
3. 5.46cm
4. 7.94cm

Page 109 Quick Test

1. a) 0.3420 b) 0.8660 c) 1
2. a) 56.5° b) 55.2° c) 88.2°

Pages 110–111 Review Questions

Page 110

1.

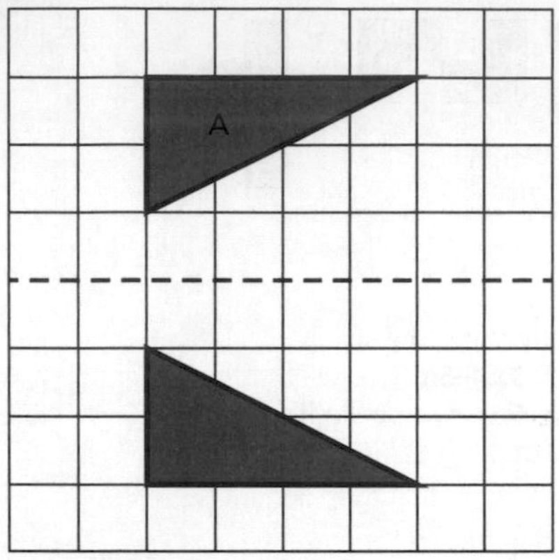

[2]

1. Order 1 **[1]**; Order 4 **[1]**; Order 2 **[1]**
2. a) See diagram [2]
b) See diagram [2]

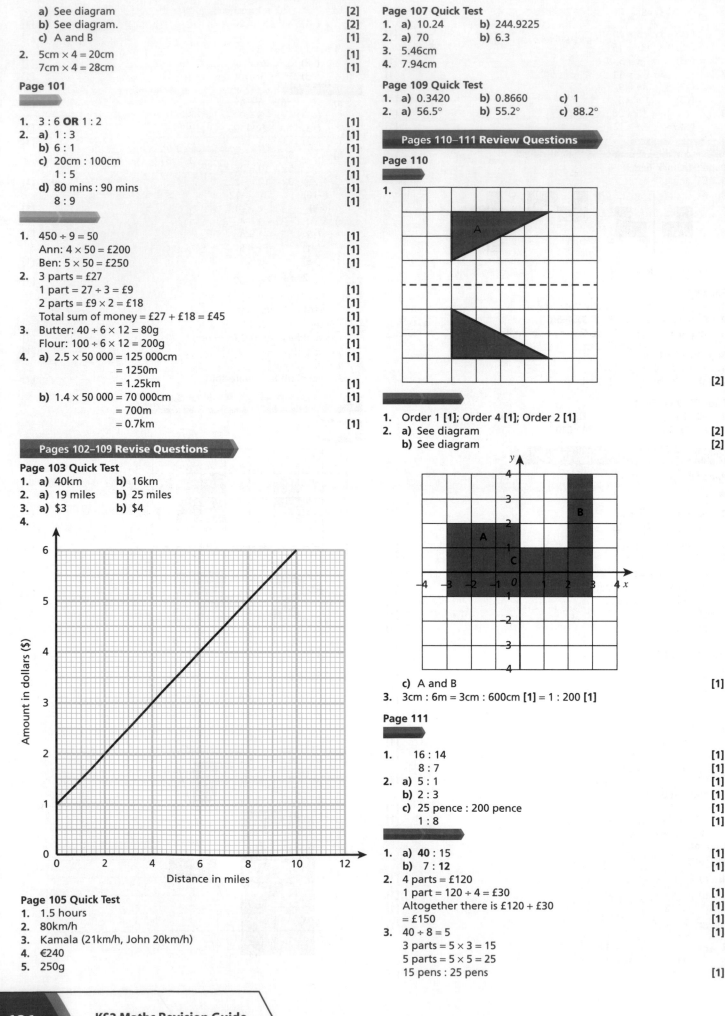

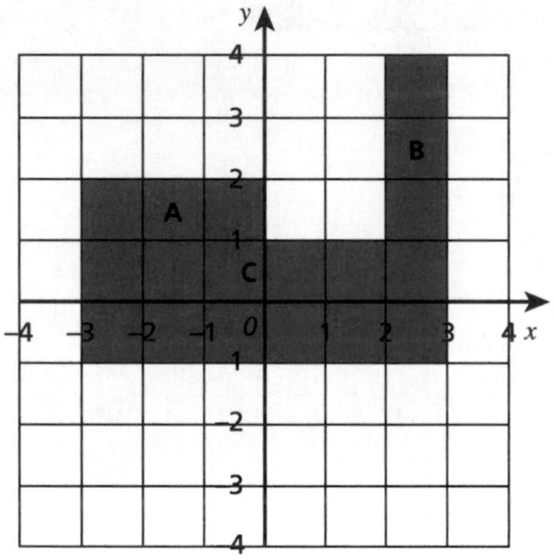

c) A and B [1]

3. 3cm : 6m = 3cm : 600cm **[1]** = 1 : 200 **[1]**

Page 111

1. 16 : 14 [1]
8 : 7 [1]
2. a) 5 : 1 [1]
b) 2 : 3 [1]
c) 25 pence : 200 pence [1]
1 : 8 [1]

1. a) **40** : 15 [1]
b) 7 : **12** [1]
2. 4 parts = £120
1 part = 120 ÷ 4 = £30 [1]
Altogether there is £120 + £30 [1]
= £150 [1]
3. 40 ÷ 8 = 5 [1]
3 parts = 5 × 3 = 15
5 parts = 5 × 5 = 25
15 pens : 25 pens [1]

4. 1 postcard = £2.16 ÷ 18 = 12 pence or £0.12 [1]
27 postcards cost £0.12 × 27 = £3.24 [1]
5. 12 ÷ 60 × 90 **[1]** = 18 foot shadow **[1]**

Pages 112–113 Practice Questions

Page 112

1. 20 × 60 [1]
= 1200mph [1]
2. 3.50 × 1.19 [1]
= €4.165 [1]

1. France, with correct reasoning, e.g. [1]
Converts \$ and € to £
1 000 000 ÷ 2.7 = £370 370 [1]
780 000 ÷ 1.54 = £506 494
OR
Converts \$ to €
1 000 000 ÷ 2.7 × 1.54 = 570 370
OR
Converts € to \$
780 000 ÷ 1.54 × 2.7 = 1 367 532
2. **a)** Speed = distance ÷ time
= 350km ÷ 1.1h [1]
= 318km/h [1]
b) Accept 2115 – 2118 (9:15 – 9:18pm) [1]

Page 113

1. **a)** 289 [1]
b) 12.25 [1]
c) 23 [1]
d) 6.4 [1]

1. **a)** $9^2 + AC^2 = 17^2$
$AC = \sqrt{289 - 81} = \sqrt{208}$ [1]
$AC = 14.4$cm [1]
b) $2^2 + 5^2 = BC^2$
$BC = \sqrt{29}$ [1]
$BC = 5.39$m [1]
2. **a)** $\sin 25° = \frac{P}{17}$
$P = \sin 25° \times 17$ [1]
$P = 7.18$m [1]
b) $\tan y° = \frac{32}{46}$
$y = \tan^{-1}(32 \div 46)$ [1]
$y = 34.8°$ [1]
3. $\cos x° = \frac{15}{22}$ [1]
$x = \cos^{-1}(15 \div 22)$ [1]
$x = 47°$ [1]

Pages 114–115 Review Questions

Page 114

1. **a)** 200 ÷ 1.75 [1]
= £114.29 [1]
b) 200 × 1.75 [1]
= US\$350 [1]
2. **a)** Time = 300 ÷ 60 [1]
Time = 5 hours [1]
b) Distance = 60 × 4 [1]
= 240 miles [1]

1. Density = 2000 ÷ 5 [1]
Density = 400kg/m³ [1]

2.

Name	Journey description
Sanjay	This person walked slowly and then ran at a constant speed.
Dee	This person walked at a constant speed but turned back for a while before continuing.
Ann	This person walked at a constant speed without stopping or turning back.
Ben	This person walked at a constant speed but stopped for a while in the middle.

All correct **[2]**; two correct **[1]**

Page 115

1. **a)** 10.89 [1]
b) 14 [1]
2. **a)** $6^2 + 8^2 = y^2$
$y = \sqrt{100}$ **[1]**
$y = 10$cm **[1]**
b) $23^2 + 15^2 = x^2$
$x = \sqrt{754}$ **[1]**
$x = 27.5$m **[1]**

1. **a)** $\tan x° = \frac{15}{8}$
$x = \tan^{-1}(15 \div 8)$ **[1]**
$x = 61.9°$ **[1]**
b) $6^2 + AB^2 = 9^2$
$AB = \sqrt{81 - 36} = \sqrt{45}$ **[1]**
$AB = 6.7$cm **[1]**
2. **a)** $8^2 + 2^2 = y^2$ **[1]** for using '2'
$y = \sqrt{68}$ **[1]**
$y = 8.2$cm **[1]**
b) Perimeter = 8.2 + 8.2 + 4 = 20.4cm (or 20.5 using unrounded values) [1]

Pages 116–127 Mixed Test-Style Questions

Pages 116–121 No Calculator Allowed

1. **a)** Surface area = 2(6 × 4 + 6 × 2 + 4 × 2) = 88cm² [1]
Volume = 6 × 4 × 2 = 48cm³ [1]
b) Surface area = 2(12 × 7 + 12 × 8 + 7 × 8) = 472cm² [1]
Volume = 12 × 7 × 8 = 672cm³ [1]
2. **a)** $4\frac{1}{2} + 2\frac{1}{3} = \frac{9}{2} + \frac{7}{3} = 6\frac{5}{6}$ [1]
b) $5\frac{2}{3} + 8\frac{1}{4} = \frac{17}{3} + \frac{33}{4} = 13\frac{11}{12}$ [1]
c) $9\frac{1}{6} - 2\frac{3}{8} = \frac{55}{6} - \frac{19}{8} = 6\frac{19}{24}$ [1]
d) $12\frac{1}{2} - 14\frac{5}{6} = \frac{25}{2} - \frac{89}{6} = -2\frac{1}{3}$ [1]
3.

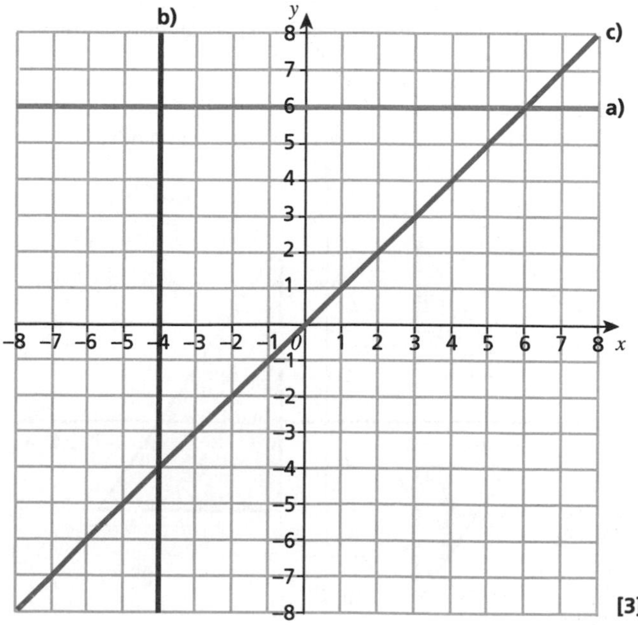

[3]

4\. a)

x	–2	–1	0	1	2	3
y	–9	–6	–3	0	3	6

[2] for all values of y correct; **[1]** for 3 or more correct.

b)

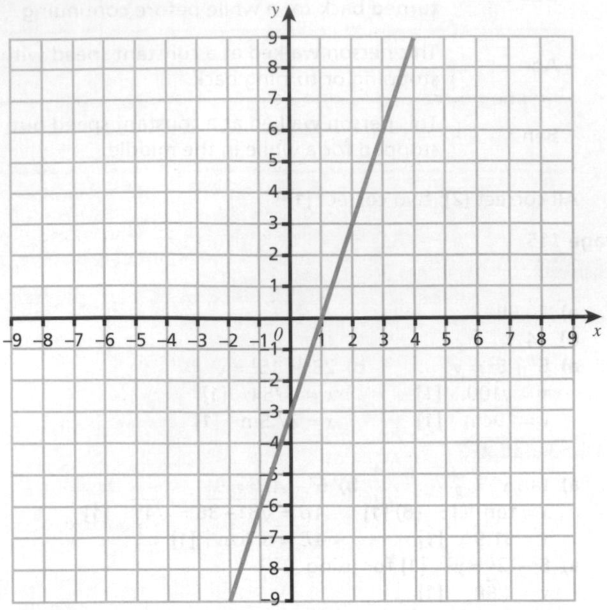

Correctly plotted line **[2]**; Straight line passing through one of the correct coordinates **[1]**

5\. a) $x = 115°$ **[1]** (corresponding and opposite), $y = 55°$ **[1]** (alternate)
b) $y = 180 - 85 = 95°$ **[1]**, $x = 180 - 75 = 105°$ **[1]**

6\. a) $4x + 4y$ **[1]**
b) $3g + 1$ **[1]**

7\. a) $4x - 20$ **[1]**
b) $4x^2 + 16x$ **[1]**

8\. a) $6(x - 2)$ **[1]**
b) $4x(x - 2)$ **[1]**

9\. $z = 2$ **[3]**; **[2]** if 9 is seen; **[1]** for a correct attempt to find the area of the trapezium with no more than one numerical error.

10\. 120 bars **[3]**; **[2]** for $\frac{7200}{60}$ or $\frac{72}{0.6}$ seen; **[1]** for 7200 or 72 seen.

> Remember to work in pounds or pence not a mixture of the two.

11\. a) 16 or 64 b) 15 or 45 or 51 or 54 or 57 or 75
c) 17 or 41 or 47 or 61 or 67 or 71 **[2]** for all three correct; **[1]** for any two correct.

12\. a) $400 + 40 = 440$mm **[1]**
b) $440 + 44 = 484$mm **[1]**

13\. a) cone **[1]** b) tetrahedron **[1]** c) cylinder **[1]**

14\.

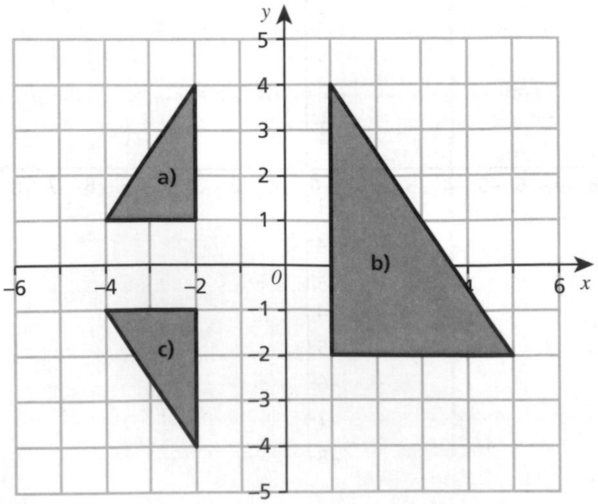

[3]

Pages 122–127 Calculator Allowed

1\. a)

Type	Frequency	Probability	
tug boat	12	$\frac{12}{40} = \frac{3}{10}$	**[1]**
ferry boat	2	$\frac{2}{40} = \frac{1}{20}$	**[1]**
sail boat	16	$\frac{16}{40} = \frac{2}{5}$	**[1]**
speed boat	10	$\frac{10}{40} = \frac{1}{4}$	**[1]**

b) $\frac{2}{5} \times 75 = 30$ **[1]**

2\. a) Surface area = 326.7cm² **[2]**; **[1]** for $8 \times \pi \times 9 + 2(4^2 \times \pi)$
Volume = 452.4cm³ **[2]**; **[1]** for $4^2 \times \pi \times 9$
b) Surface area = 298.5cm² **[2]**; **[1]** for $10 \times \pi \times 4.5 + 2(5^2 \times \pi)$
Volume = 353.4cm³ **[2]**; **[1]** for $5^2 \times \pi \times 4.5$

3\. 3.43cm² (to 2 d.p.) **[3]**; **[2]** for 16 **and** $\pi(2^2)$ or 12.566 37 seen; **[1]** for 16 **or** $\pi(2^2)$ or 12.566 37 seen.

> Find the area of the square and subtract the area of the circle. Remember the diameter of the circle is the same as the side length of the square, in this case 4cm.

4\. Carol's cars and £6000 and £6050 seen **[3]**; **[2]** for £6000 and £6050 seen but no conclusion; **[1]** £5400 seen.

5\. a) 8.333m/s **[2]**; **[1]** for 100 ÷ 12 seen.
b) 30km/h **[2]**; **[1]** for 0.1km ÷ 0.003 333h seen.

6\. 10.6m **[2]**; **[1]** for sin 45° × 15

7\. 83.6km (±1km) **[3]**; **[2]** for 20km + 22.4km + 41.2km seen; **[1]** for 2 correct measurements in km.

8\. a) 0.08 **[1]**
b) $\frac{2}{25}$ **[1]**

9\. a) 160g of butter **[1]**, 400g of flour **[1]**
b) 200g of butter **[1]**, 500g of flour **[1]**

10\. a) 7 **[1]** b) 16 **[1]**

11\. a)

1	0	0	4	5	
2	0	3	5	8	8
3	4	5	9		
4	5	8	9		

Key: 2|0 = 20 minutes
Correct diagram with key **[2]**; correct diagram only **[1]**
b) 28 minutes **[1]**

Glossary

a

alternate angles which are created on a set of parallel lines are the same on a 'Z' angle.

angle the space (usually measured in degrees) between two intersecting lines or surfaces at or close to the point where they meet.

area the space inside a 2D shape.

arithmetic sequence a sequence of numbers with a common difference.

axis a line that provides scale on a graph. Often referred to as x-axis (horizontal) and y-axis (vertical).

b

biased a statistical event where the outcomes are not equally likely.

bisect to cut exactly in two.

brackets symbols used to enclose a sum.

c

centre of enlargement the position from which the enlargement of a shape will take place.

centre of rotation the point about which a shape is rotated.

certain an outcome of an event which must happen, probability equals 1.

chart a visual display of data.

circle a round 2D shape.

circumference the perimeter of a circle.

class interval the width of the group (difference between the upper and lower limit of the group).

composite or compound a complex 2D or 3D shape made from several simpler shapes.

congruent exactly the same.

constant a value that does not change.

conversion to change from one unit to another.

coordinates usually given as (x, y); the x-value is the position horizontally, the y-value the position vertically.

correlation the relationship between data, the 'pattern'. Can be positive or negative.

corresponding angles which are created on a set of parallel lines are the same on an 'F' angle.

cos (cosine) the ratio of the adjacent side to the hypotenuse in a right-angled triangle.

cylinder a 3D shape with a circular top and base of the same size.

d

data a collection of answers or values linked to a question or subject.

decimal a number that contains tenths, hundredths, etc.

decimal places the number of places after the decimal point.

decimal point a point used to separate the whole part of a number from the fraction part.

decrease to make smaller.

degree the unit of measure of an angle.

denominator the bottom number of a fraction.

density the mass of something per unit of volume.

diameter the distance across a circle, going through the centre.

difference subtraction.

direct proportion quantities are in direct proportion if their ratio stays the same as the quantities increase or decrease.

distance length.

divide to share.

double multiply by 2.

e

edge a line where two faces meet in a 3D shape.

enlargement a shape made bigger or smaller.

equation a mathematical statement containing an equals sign.

equivalent the same as.

estimate a simplified calculation (not exact), often rounding to 1 significant figure.

even chance an equally likely chance of an event happening or not happening.

event a set of possible outcomes from a particular experiment.

expand remove brackets by multiplying.

experimental probability the ratio of the number of times an event happens to the total number of trials.

expression a collection of algebraic terms.

exterior angle an angle outside a polygon formed between one side and the adjacent side extended.

f

face a side of a 3D shape.

factor a number that divides exactly into another number.

factorise take out the highest common factor and add brackets.

formula a rule linking two or more variables.

fraction any part of a number or 'whole'.

frequency the number of times 'something' occurs.

function machine a flow diagram which shows the order in which operations should be carried out.

g

gradient the measure of steepness of a line.

graph a diagram used to display information.

grouped data data which has been sorted into groups.

h

highest common factor the highest factor two or more numbers have in common.

hypotenuse the longest side of a right-angled triangle.

hypothesis a prediction of an experiment or outcome.

i

impossible an outcome of an event which cannot happen, probability equals 0.

improper fraction a fraction where the numerator is larger than the denominator.

increase to make bigger.

index the power to which a number is raised. In 2^4 the base is 2 and the index is 4.

integer whole number.

intercept the point at which a graph crosses the y-axis.

interest an amount added on or taken off.

interior angle the measure of an angle inside a shape.

interpret to describe the trends shown in a statistical diagram or statistical measure; the way in which a representation of information is used or surmised.

inverse the opposite of.

k

key a statement or code to explain a mathematical diagram.

l

likely a word used to describe a probability which is between evens and certain on a probability scale.

line of best fit the straight line (usually on a scatter graph) that represents the closest possible line to each point; shows the trend of the relationship.

linear in one direction, straight.

lowest common multiple the lowest multiple two or more numbers have in common.

lowest terms a simplified answer.

m

mean a measure of average; sum of all the values divided by the number of values.

median a measure of average; the middle value when data is ordered.

mixed number a number with a whole part and a fraction.

mode a measure of average; the most common.

mutually exclusive events that have no outcomes in common.

n

negative below zero.

net a 2D representation of a 3D shape, i.e. a 3D shape 'unfolded'.

***n*th term** see *position to term.*

numerator the top number of a fraction.

o

ordinary number a number not written in standard form.

outlier a statistical value which does not fit with the rest of the data.

p

parallel lines are said to be parallel when they are at the same angle to one another, and never meet.

parallelogram a quadrilateral with two pairs of equal and opposite parallel sides.

percentage out of 100.

perimeter distance around the outside of a 2D shape.

perpendicular at 90° to.

pi (π) the ratio between the diameter of a circle and its circumference, approx. 3.142.

pictogram a frequency diagram in which a picture or symbol is used to represent a particular frequency.

pie chart a circular diagram divided into sectors to represent data, where the angle at the centre is proportional to the frequency.

place value indicates the value of the digit depending on its position in the number.

position to term a rule which describes how to find a term from its position in a sequence.

positive greater than zero.

power see *index*.

prime a number with exactly two factors, itself and 1.

prism a 3D shape with uniform cross-section.

probability the likeliness of an outcome happening in a given event.

probability scale a scale to measure how likely something is to happen, running from 0 (impossible) to 1 (certain).

product multiplication.

protractor a piece of equipment used to measure angles.

Pythagoras' Theorem in a right-angled triangle, the square on the hypotenuse is equal to the sum of the squares of the other two sides.

q

quadratic based on square numbers.

quadratic equation an equation where the highest power of x is x^2.

quadrilateral a four-sided 2D shape.

quantity an amount.

r

radius half the diameter; the measurement from the centre of a circle to the edge.

range the difference between the biggest and smallest number in a set of data.

ratio a comparison of two amounts.

raw data original data as collected.

ray a line connecting corresponding vertices.

reflection a mirror image.

regular polygon a 2D shape that has equal-length sides and angles.

rotation a turn.

rounding a number can be rounded (approximated) by writing it to a given number of decimal places or significant figures.

s

sample space a way in which the outcomes of an event are shown.

scale the ratio between two or more quantities.

scale factor the number by which a shape/number has been increased or decreased.

scatter graph paired observations plotted on a 2D graph.

sector a section of a circle enclosed between an arc and two radii (a pie piece).

sequence a set of numbers or shapes which follow a given rule or pattern.

share to divide.

significant figures the importance of digits in a number relative to their position; in 3456 the two most significant figures are 3 and 4.

similar two shapes that have the same shape but not the same size.

simplify make simpler, normally by cancelling a fraction or ratio or by collecting like terms.

simultaneous equations equations that represent lines that intersect.

sin (sine) the ratio of the opposite side to the hypotenuse in a right-angled triangle.

solve work out the value of.

speed how fast something is moving.

square a regular four-sided polygon; to multiply by itself.

square number a number made from multiplying an integer by itself.

square root the opposite of squaring; a number when multiplied by itself gives the original number.

standard form a way of writing a large or small number using powers of 10, e.g. $120\,000 = 1.2 \times 10^5$.

substitute to replace a letter in an expression with a number.

sum addition or total.

surface area the total area of all the faces of a 3D shape.

survey a set of questions used to collect information or data.

t

tan (tangent) the ratio of the opposite side to the adjacent side in a right-angled triangle.

term to term the rule which describes how to move between consecutive terms.

tessellation a pattern made by repeating 2D shapes with no overlap or gap.

trapezium a quadrilateral with just one pair of parallel sides.

triangle a three-sided 2D shape.

u

unlikely a word used to describe a probability which is between evens and impossible on a probability scale.

units these define length, speed, time, volume, etc.

v

vertex the point where the edges meet on a 3D shape.

volume the capacity, or space, inside a 3D shape.

Index

addition 32, 55, 57
algebra 34–7, 41, 51
alternate angle 68
angles 46, 66–9, 76, 86, 109
area 18–21, 28, 38
arithmetic sequences 10–11
averages 22
axis 47, 58, 102

biased event 70
BIDMAS 7
bisecting 67
brackets 36, 84

calculating probability 72
centre of enlargement 91
centre of rotation 90
certain probability 70
circles 21, 44
circumference 21
class interval 24
comparison 47
composite shapes 45
compound shapes 19
congruence 92
constant speed 104
conversion graphs 102–3
coordinates 58–61, 65, 75
correlation 49
corresponding angle 68
cos 108
cuboids 43
cylinders 44

data 24–5, 29, 39, 46
decimal places 33
decimal point 30, 32
decimals 30–3, 40, 50, 78, 88, 98
decrease 80
degrees 66
denominators 54, 78
density 105
diagrams 48
diameter 21, 44
difference 22
direct proportion 97
distance 104
division 7, 30, 33, 56
doubling 85

edges 42
enlargement 91–3, 100, 110
equations 34, 60–1, 82–5, 89, 99
equivalent fractions 33, 54
equivalent ratio 95
estimating 33
even chance 70
event 70, 71, 72
expanding 36
experimental probability 73
expressions 34–5
exterior angle 69

faces 42
factorising 37
factors 8, 95
formulas 35
fractions 33, 54–7, 64, 74, 78, 88, 98
frequency chart 23
frequency diagrams 47
function machine 10

gradient 59, 60
graphs 48, 58–61, 65, 75, 102
grouping data 24–5

highest common factor (HCF) 9
hypotenuse 106
hypotheses 49

impossible probability 70
improper fractions 56–7
increase 80
indices 30
integers 6
intercept 59, 60
interest 81
interior angle 69
interpreting data 25, 46–9, 53, 63
inverse 56, 82

keys 24

likely probability 70
line of best fit 49
linear graphs 59, 60
longest side 106
lowest common multiple (LCM) 9
lowest terms 95

mean 22
median 22
mixed numbers 56, 57
mode 22
multiplication 7, 30, 32, 56
mutually exclusive event 72

negative numbers 6, 84
nets 42
*n*th term 12–13
numerator 54, 78

outliers 22

parallel lines 20, 68
parallelograms 20
percentage 46, 78, 79, 80–1, 88, 98
perimeter 18, 19, 28, 38
perpendicular height 18

perpendicular line 90
pictograms 46
pie charts 46, 48
place value 6
polygons 68–9
position to term rule 12
positive numbers 6
powers 30
prime factors 8
probability 70–3, 77, 87
probability scale 70
product 8, 35
proportion 94, 97, 101, 111
protractor 66
Pythagoras' Theorem 106

quadratic equation 61
quadratic graphs 61
quadratic sequences 13
quadrilaterals 67
quantities 78–9, 80–1

radius 21, 44
range 22
rates 104–5, 112, 114
ratio 93, 94–6, 101, 108, 111
raw data 24
rays 91
real-life graphs 102, 112, 114
rectangles 18
reflection 90
right-angled triangles 106–9, 113, 115
rotational symmetry 90
rounding 33

sample spaces 71
sector 21
scale drawing 92, 93
scale factor 91
scatter graphs 48–9
sequences 10–11, 13, 27
shortest side 107
significant figures 33
similarity 91, 92
simplifying 34, 95
simultaneous equation 61
sin 108
solving equations 83, 84, 85
speed 104
square numbers 8, 106
square roots 8, 106
standard form 30
statistics 22, 29, 39
stem-and-leaf diagrams 24
substitution 12, 35, 61
subtraction 32, 55, 57
sum 22
surface area 43, 44, 52, 62
surveys 49
symbols 6
symmetry 90, 100, 110

tabulating events 72
tally charts 23
tan 108
term to term rule 11
tessellation 69
3D shapes 42–5, 52, 62
time graphs 104
translation 90
trapeziums 20
triangles 18, 66, 106–9
two-way tables 25

unitary method 97
unit pricing 105
units 93, 97
unknown numbers 82, 83
unlikely probability 70

vertices 42
volume 43, 44, 45, 52, 62

x-axis 58

y-axis 58

Notes